Poultry
SIGNALS®

LAYING HENS

A PRACTICAL GUIDE FOR LAYER FOCUSED MANAGEMENT

ROODBONT
PUBLISHERS

LOUIS BOLK
INSTITUUT

LIVESTOCK RESEARCH
WAGENINGEN UR

GD

Credits

Poultry Signals: Laying Hens

Authors original edition Poultry Signals

Monique Bestman
Marko Ruis
Jos Heijmans
Koos van Middelkoop

Content Editors Poultry Signals, edition Laying Hens

Wouter Steenhuisen
Theo Peters
Phill te Winkel
Koos van Middelkoop

Final editing

Ton van Schie

English Translation

Agrolingua

Photography

Photography Cover
Front: Marcel Bekken (t), ASG (b)
Front flap: LBI
Back: ASG
Back flap: LBI , ASG, GD, Koos van Middelkoop

Photography Interior:
Alpharma (116), Andries de Vries (15, 16), Arthur Slaats (28, 48), ASG (10, 16, 18, 24, 25, 26, 27, 29, 31, 35, 36, 38, 38, 40, 41, 49, 54, 69, 73, 75, 75, 76, 80, 81, 85, 87, 88, 91, 92, 93, 95, 96, 100, 102, 103, 104, 107, 109, 118, 119), AviVet: Roland Bronneberg (92), Back yard farming (50), Bastiaan Meerburg (103), BiotechMichael (61), Bloemendaal Eierhandel (94), C. Bennet(47), Christel Lubbers (98), Edward Mailyan (91), GD (57, 28, 50, 60, 64, 83, 89, 91, 92, 101, 105, 106, 107, 108, 108, 110, 111, 112, 113, 114, 115, 116, 117, 118, 63), GULLI.ver (15), Henk Heidekamp (53), Henk Rodenboog (50), Hilly Speelman (42), Interbroed leghennen (66, 67, 80), internet (98), Jansen Poultry Equipment (24), Koos van Middelkoop (6, 10, 15, 16, 32, 33, 47, 50, 51, 69, 70, 71, 72, 79, 88, 89, 90, 91, 93, 95), LBI (4, 6, 8, 9, 17, 18, 22, 25, 28, 29, 32, 42, 43, 44, 46, 48, 56, 58, 60, 68, 74, 75, 78, 84, 86, 87, 90, 98, 101, 108), Marcel Berendsen (8, 9, 11, 14, 25, 26, 30, 31, 33, 40, 48, 49, 51, 54, 55, 57, 58, 60, 62, 63, 64, 64, 73, 75, 76, 77, 86, 92, 93, 100, 107), Moba (14, 93, 95, 96, 97), Pas Reform (45, 46, 47, 51, 91, 93, 99), PTC+, Helmich van Rees (7, 32, 33, 45, 52, 62, 79, 96, 98), Rob van Veldhuizen (103), Ron Jöerissen (55), Schippers BVBA (109), Silly Chick (26), Tine Jansen (46), Ton van Schie (94), Twinpack (98), Vencomatic (5, 6, 7, 9, 12, 15, 16, 24, 25, 26, 29, 31, 39, 43, 57, 94), Verbeek (104), Wayne Skews, poultry-farming.co.za (114), Wouter Steenhuisen (46, 91, 91, 96, 109), Zonne-Ei-Farm B.V (5).

Illustrations

Marinette Hoogendoorn

Design

Dick Rietveld, Erik de Bruin, Varwig Design

Advisor

Jan Hulsen, Vetvice Groep

Special thanks to:

Peter van Agt, Marleen Boerjan, Pieter Bouw, Mijndert van den Brink, Hilko Ellen, Rick van Emous, Marrit van Engen, Teun Fabri, Thea Fiks, Niels Geraerts, Arjan Gussinklo, Jan van Harn, Wim Hoeve, Jan Hulsen, Ron Jöerissen, Ingrid de Jong, René Kieftenbelt, Gerjan Klok, Cécile Korevaar, Marinus van Krimpen, Pieter Kruit, Jan en Marcel Kuijpers, Ferry Leenstra, Sander Lourens, Jac Matijsen, Monique Mul, Bert van Nijhuis, Kees van Ooijen, Wim Peters, Dr. David Pollock, Bianca Reindsen, Berry Reuvekamp, Henk Rodenboog, Jorine Rommers, Piet Simons, Arthur Slaats, Alex Spieker, André van Straaten, Otto van Tuil, Cor van de Ven, Jan-Paul Wagenaar, Ruud van Wee, Sible Westendorp, Helmich van Rees, Laura Star, Joost Koster, Jeroen van der Heijden, Paul Buisman, Henry Arts, Gerd de Lange, Merel van der Werf, Mari van Gruijthuijzen, Richard Wentzel, Andries de Vries, Jacco Wagelaar, Jan Dirk van der Klis en Karin Jonkers.

Roodbont Publishers B.V.
P.O. Box 4103
7200 BC Zutphen
The Netherlands
T +31 (0)575 54 56 88
info@roodbont.com
www.roodbont.com

Louis Blok Instituut
www.louisbolk.org

Livestock Research Wageningen UR
www.livestockresearch.wur.nl

GD Deventer
www.gddeventer.com

This publication has been made possible with the support of the Dutch province Gelderland.

© Roodbont Publishers B.V., 2012

Laying Hens is part of the Poultry Signals© book series.

ISBN 978-90-8740-124-5

Contents

Focus on chickens

Poultry keeping is all about the chickens. In addition to working efficiently and neatly, providing a comfortable environment for your birds and looking after them properly is a basic requirement. Laying Hens is about how to house and manage chickens in the best possible way, so the focus is always on the chicken. What is a chicken and what are its needs in terms of health, welfare and production? A good poultry farmer looks after his birds attentively and does this several times a day. Not only during the day when it is light, but also in the evening when it is dark, making sure he covers the whole flock. He checks the birds, he knows the poultry house and he responds promptly to irregularities and problems, and he also has many well defined routines to help him. A poultry farmer not only needs to know what a healthy flock should look like, but he must also be able to notice that his birds are OK by looking at them.

Observing or understanding?

Everyone can learn how to observe chickens properly and understand them so that they can manage their birds' health, welfare and production better. Some people have a natural ability to observe and understand chickens. Others have to go through a lot of trouble to learn. But everyone improves with practice. Make a note of anything unusual you see and what you do about it.

Many people find that when it comes to observing and assessing their birds they can't see the wood for the trees. They are so involved in the farm that they no longer notice irregularities because they are so used to seeing them. So identify your own blind spots and eliminate them. Open your mind to others and to new things. Be critical. And don't be afraid to change.

An important question a good entrepreneur regularly asks himself is: am I getting the very best out of everything? Or could I be getting even more?

Changing situations

Table egg production is in a state of flux: in some countries cage systems are being phased out, while in other countries cages are being subjected to ever more stringent requirements. So many table egg producers in Europe will have to make a choice when making alterations or building new facilities: should I go for an enriched cage system or an aviary system? Do I go free-range or even organic? The choice of system is up to the poultry farmer, and will depend on regulations, personal preference, financial returns and the environmental options he has. A new system places different demands on the poultry farmer. The more opportunity chickens get to express their natural behaviour, the more you can read them. However, such alternative systems demand better skills of the poultry farmer to be able to keep the birds in good health.

"Laying Hens hones your ability to pick up signals from chickens and use them to manage and improve them."

Train yourself to look and see

Proper management of your birds begins with critical observation. Look with awareness. Not only at the birds themselves but also at the manure and the eggs. Take a step back (sometimes literally) and take your time. You can't look at things with awareness if you are doing something else at the same time. Stop and think about the signals your birds are sending. The longer you spend on this, the more subtle signals you will pick up. It takes skill and insight to see signals before the consequences reveal themselves.

The red thread running through this book is 'look, think and act'. The three basic questions a poultry farmer must keep on asking themselves are:

1. **What am I observing?**
2. **Why is this happening?**
3. **What should I do?**

Not a manual

Laying Hens is not a manual listing standards for all aspects of poultry farming. It is a guide that will help you learn to look after your birds properly and translate their signals into actions that will help you run your business better. The book is written in a form that we hope will invite you to dip into it regularly, giving you lots of new ideas each time.

Production - housing and care - health

Production, housing/care and health are as firmly interlinked as the sides of a triangle.

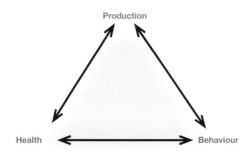

The one affects the other. Production is the central goal of any farmer or entrepreneur. Production is closely related to housing, care and health. You can not only tell a lot about the health of your flock by watching the birds' behaviour and production, but also about the quality of the housing and care. So you can always follow the triangle from different starting points.

Top athletes

The production of eggs is a top athletic achievement for the birds. It goes without saying that there is a big difference between layers in different circumstances. The poultry in different subsectors and even different systems can be compared with different kinds of athletes. Athletes also come in all shapes and sizes depending on their discipline. The chicken, the type of housing and the farmer must all suit one another.

Layers: less and less difference

Twenty years ago layers could be classed as light-weight and medium-weight. Light-weight hens were usually white feathered and laid white eggs, and medium-weight birds were brown and laid brown eggs. White hens often laid slightly more eggs, needed less food and were more flighty, active or nervous. There is still a difference between white and brown, but the birds have since become more alike in terms of weight, although not in behaviour or production performance. They are still very different in this regard. Both types are kept in cage and floor systems. Medium-weight Silvers have also made an appearance in recent years. These birds have white feathers interspersed with the occasional brown one. They lay brown eggs and have the weight and nature of brown hens.

Laying hen in a cage system: the stayer

Like long-distance runners, this type of bird needs to perform for a long time, so it needs stamina and must be in good condition.

Laying hen in alternative systems: the steeple-chaser

Floor layers are faced with different challenges from caged birds. So their lives can be compared with a steeple-chase.

Brown hens are often slightly heavier than white ones so they need about 10% more feed for maintenance.

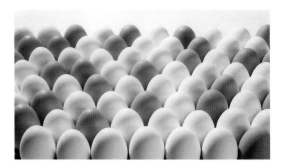

Silver hens are also often referred to as 'white on the outside, brown on the inside' because of their white appearance and other characteristics of brown breeds.

The egg colour is the starting point when choosing a chicken. Next up are her production characteristics, behaviour and feather colour.

Alternative breeds

Although only a small number of breeds of laying hens are used worldwide, some breeds have very specific characteristics that deliver benefits in certain conditions. One example is the Kuroiler (see photo). These birds are resistant to certain diseases and are more robust, but less efficient. Where housing or management conditions are poor, this can be an advantage, especially when the birds have to search for food themselves. In commercial operations, feed costs play a key role in the financial results, so it is better to go for an efficient breed. The Egyptian Fayoumi is another example.

Key factor: the poultry farmer

If you know what your strengths and weaknesses are, you can exploit your strengths and make sure your weaknesses are covered in other ways.

The hands-on person is fully focused on getting the work done. He tries to make conditions as pleasant as possible. But does he also pick up on problems with the birds in time?

The poultry keeper gets most satisfaction from working among the birds, but does he notice that there is other work that needs doing?

The entrepreneur/manager concentrates on organising and running the business. But does he spend enough time in the poultry house?

Many farms have staff. With caged systems it is easier to use hired or unskilled labour. A higher level of management in the poultry house is needed for floor systems, as you need to be able to pick up and deal with signals straight away. This system is therefore not only more labour-intensive but also needs a different type of worker.

CHAPTER 1

Seeing more by looking more closely

If you only look at technical aspects such as laying rates, egg weight, mortality and egg quality, growth or feed and water consumption, you run the risk of missing important signals and being overtaken by events. You can pick up these signals from the chickens themselves and from their appearance, behaviour, manure and eggs. Use the chicken as an 'informant'.

In a nice healthy flock you want to identify irregularities as soon as possible so as to avoid problems.

Use all your senses. Even before you enter the poultry house, you'll hear whether the hens sound different. Stand outside the door for a moment; don't enter straight away. When you enter, you'll smell whether there is a problem with the ventilation. Use your eyes and ears to see and hear how active the birds are and whether they are reacting differently to your presence than usual. Also use your senses to observe the heat and cold in the poultry house. Every irregularity needs to be attended to. A different smell may, for example, indicate that the birds have not had any feed the previous day.

Farm blindness

You can only recognise irregularities if you know what is normal. You will learn what is normal by

observing as often and as objectively as possible. But be aware of the dangers of farm blindness. Farm blindness is when you see the situation on your own farm as the norm. Limit farm blindness by talking to colleagues and advisers. Act on critical comments.

Limit farm blindness by being critical and talking to colleagues and advisers.

Picking up the signals

You can pick up a lot from a flock while you are sweeping the paths, collecting eggs, picking floor eggs in a floor system, and scattering grain. But make sure you also do some inspections without doing other things at the same time. You will be more aware of the signals because you are giving the hens and their environment/the system your full attention. Activities - but also moods - detract from your observations. You will also disrupt the hens' behaviour, causing relevant signals not to be displayed or to be lost. The chickens will also react less to your presence if you enter the house more often without doing anything specific. An inspection is a tour of the whole house, not just the front or just one aisle. Make sure that you not only pay attention to the birds but also to the drinking water and the feeding system. So take a look everywhere: front, back, middle, bottom, and do not forget to look at the top. This applies to both cages and aviary systems.

Look from flock to chicken

Start your inspection by observing the whole flock. In a cage system, check whether the eggs are evenly distributed throughout the system and make sure there isn't a pile of eggs somewhere. Also pick up the chicken that is always standing at the back of the cage, and have a good look at her and feel her.

In a floor system it is important that the animals are spread out all over the space. Are they making use of the different parts of the house? Are they avoiding certain places, maybe because the climate is bad there (draught, cold)? Try to spot the differences between birds. Are they uniform? How do they differ? In alertness, condition or in another way? Pick up birds that seem different and take a closer look. If you discover an irregularity, see whether it is an incidental case or a signal of a bigger underlying problem. Also pick up some birds at random and assess them. Irregularities are not always immediately evident.

Things you observe in detail can only be properly assessed in their context. So look from chicken to flock as well. Sometimes you will need to take a step back to be able to see things better.

Go and stand in the house for fifteen minutes or so now and again and look at the birds quietly. Or put a chair in the house and sit on it for a few minutes at regular intervals. Only then will you pick up on any irregular behaviour.

This hen is not healthy and is a source of trouble. Take this animal out of the flock.

Use all your senses. Listen at the door before you enter the poultry house. As soon as you go in the hens will start reacting to you, so you won't hear the sound they make when they are quiet. Also, don't turn up the light straight away to see better as this will also influence their behaviour.

Using the signals

Use what you see to improve your flock management. Ask the following questions about everything you see:

1. **What am I seeing (hearing, smelling, feeling)? What is the signal?**
2. **Why is this happening? What is the explanation?**
3. **What should I do? Can I leave it or should I take action?**

A genuine signal will be repeated. Think about what you are seeing and how it relates to the circumstances: does it happen often? At different times? To different birds? On other farms? Go and see for yourself or ask people. Also go and look in the evening and at night.

Know when risks are likely to occur and keep one step ahead by eliminating them or being extra alert if you are expecting them.

Risk birds

There will always be some risk birds in a flock, such as poorly developed ones. They will be the first to suffer from disease, lack of water or other shortages. These are the signal birds. The risk birds are the first ones to tell you that something is wrong. All the more reason to be alert to them. Risk birds also include those whose behaviour or appearance could cause problems. Not as a victim but as the cause. Think about which birds and problems they are in terms of your particular farm and how you can respond to them promptly.

Frightened or sick hens in the nests will soil the nests and the eggs.

How to observe your birds in a structured way:

1. Look at them both with and without doing other things.
2. Look at the whole flock, the individual hen and then back to the flock.
3. Look for averages and extremes.
4. Look at the front, back and middle of the poultry house. The same applies to processes such as feeding. Look at the front, middle and back of the feeder that is being filled. What is happening there?
5. Look at different times and in different circumstances.
6. At set times, stand still in the poultry house; don't keep walking round the whole time.
7. Identify risk times, risk birds and risk places.

Avoid overcrowding. It is best to assess this when the chain feeder is running. All the hens must be able to eat. If all the birds in a cage can't eat next to each other, there is a problem with the cage type and its bird density. With floor systems, birds constantly running to and fro is a signal of overcrowding.

Risk times

There are certain times of the day or season or certain times during an inspection that can be risky. Known, recurring risk times are the feeding routines. Make sure your feeding machine and its weighing scale are working properly.

Risk times can also last for several days or weeks. The time when young hens come into lay is a risk time, for example. This is not only a risk time for the birds themselves but also for the operation of your system. If you open the nests too soon, they can become contaminated. If you open them too late, the hens will lay outside the nests. During winter there is a risk of too little ventilation. If you are using a floor system, you have to make more effort to keep the house climate and the litter in optimal condition in winter. Make sure the climate is OK at bird level and not only for you as a farmer. In summer layers can suffer from heat stress.

Risk places

In every poultry house there are risk places where you can expect problems. Places you know could pose a risk should be a permanent part of your daily s. So make an effort not to miss them. There are risk places in both floor and cage systems. In the latter case, that has more to do with the position in the poultry house. Cages under the air inlet are often draughty, so the birds don't look good. More light causes more pecking and consequently poorer feathering.

Unclassified notable observations

Sometimes you encounter things you don't immediately understand. Of course not every unclear signal will necessarily cause harm. A term for this: unclassified notable observations (UNO, 'you know). When you do such observations, you need to find out why they occur. You will learn most by trying to understand how good situations come about: in other words, what the success factors are.

> ### Related signals
> An indicator of the distribution of birds on perches at night is the amount of manure you find under the slats or on the various manure belts. If the distribution is uneven then the hens are roosting unevenly.

This young hen has her head pulled in. Rearing is a risk period in the life of a hen: it should be using all its energy to have a good start. This hen can't and will be one of the first victims in case of a disease.
Do you see several hens like this? It might be a signal that there is something wrong with your management.

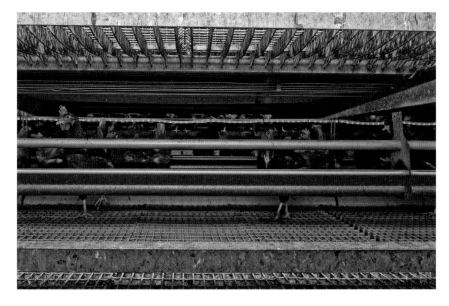

On this farm, the nests are suspended on the external wall. It is dark under the nests: a risk place for floor eggs. Solution: some extra lighting has been installed.

Farm records show objective signals

Valuable signals can also be obtained from farm records. Farm recording is an active process of collecting, processing and analysing information in order to improve results.

The results obtained after the processing of data can be divided into two categories. The first category are the technical results e.g. mortality percentage, feed conversion etc. The second category consists of the financial results and is expressed in terms of costs and revenues.

Use the information you collect. By collection data at the same time of the day, you will notice any irregularities sooner. Clear changes in water consumption are the first signals of health problems. But they can also be caused by the water or feed supply itself. Changes in feed consumption are also a signal (feed not uniform, demixing of feed in de feeding system?).

The cycle of farm recording

Keeping records should help you constantly improve your results.

→ flock looks fine

→ good condition

→ feathering looks good

No further remarks

Ask your advisers to record their observations in your logbook.

1. Collect
Obtain and calculate key figures during and at the end of the production cycle

2. Analyse
Analyse/compare the obtained results with standards/earlier production cycles/earlier weeks within production cycle/other farms in the region.

3. Improve
Make and implement a plan to improve technical and/or financial results.

Record the data where you collect it in order to prevent loss of information. In this case: in the poultry house.

> **Kind of information: production, water, feed, health, financial, etc.**
>
> **Frequency:** daily, weekly, periodically, etc.
> **Responsibility:** the farm worker, the farmer, the production supervisor or the manager.
> **Type of information:** production control, final flock results, health supervision, planning.

Most important data

These are minimal figures you need to know how a flock is doing. You should collect data for at least four weeks to judge weekly averages (the figures will fluctuate daily).

Mortality

Weekly mortality percentage (%)

$$\frac{\text{number of dead layers this week}}{\text{number of layers housed}} \times 100\ \%$$

Cumulative mortality percentage (%)

$$\frac{\text{Cumulative number of dead layers}}{\text{number of day-old layers housed}} \times 100\ \%$$

General targets
White layers: 0.7-0.8% per 4 weeks. Total mortality 9-10% (64 weeks).
Brown layers: 0.5-0.6% per 4 weeks. Total mortality 6.5-9% (64 weeks).

Points for attention
Mortality is a bit higher at start of lay until peak production. It also increases slightly towards the end. ('burnt out hens'). The average should not be more that 1% per month. If mortality increases by 0.5% or more within a week, there is something wrong. Especially if this happens a few weeks in a row.

Production

Number of hen days (hd):
(# hens at beginning of a period + # hens at end of a period)/2 x no of days in period

Laying-rate (%)

$$\frac{\text{number of eggs collected}}{\text{number of hen days}} \times 100$$

Cumulative number of eggs per hen housed (eggs/hh)

$$\frac{\text{cumulative number of eggs collected}}{\text{number of hens housed}}$$

Can also be cumulative egg mass per hen (kg egg/hh), in case payment is per kg egg

General targets
White layers: 280-330 eggs per production period (64 wks). Average laying % hen day: 75-80%. Egg weight: 60-62 g.
Brown layers: 275-325 eggs per production period (64 wks). Average laying % hen day: 71-79%. Egg weight: 62-64 g.

Peak production: at 5-10 weeks after start of lay, the laying percentage is 92-95%. After 10 months of production the laying percentage will be around 70%.
Kg of eggs per hen housed (hh) is 18-19 kg per production period of one year.

Points for attention
At start of lay, the laying percentage should double every week (i.e. 8%-16%-32%-64%) towards the peak. After peak production, you should be concerned if the laying percentage drops by more than 4% per week.
The percentage of second grade eggs over the whole production period should be 2-5%.

Egg weight should increase weekly from the first egg of 48 g to about 60 g at age 30 weeks and 65-70 g at the end. It may then fluctuate by 0.5 g per week: any more than that is an important signal. The egg weight at 40 weeks is a good indication of the average egg weight over the whole production period. High environmental temperatures have a negative influence on the egg weight.

The number of eggs per hen housed is a better measure than the number of eggs per hen day (hd), since it also includes the effect of mortality on results.

Feed efficiency

Feed consumption per layer per day (gram)

$$\frac{\text{kg feed in period}}{\text{number of hen days}} \times 1000$$

Feed Conversion Ratio (FCR)

$$\frac{\text{total amount of feed in a period}}{\text{net weight eggs collected in the same period}}$$

Grams of feed per egg (gram/egg)

$$\frac{\text{total amount of feed in a period in grams}}{\text{number of eggs collected during the same period}}$$

General targets
White layer: 105-115 gram per day (42 kg per hen/laying period)
Brown layer: 115-125 gram per day (45 kg per hen/laying period)
(based on feed with 2800 kcal ME per kg).
Feed Conversion Ratio (FCR) indicates how efficiently the hen uses the feed; this should be 1.90-2.50.

Points for attention
Body weight is a good indication of feed consumption and should always increase. It should increase substantially before peak production and less afterwards, but it should never decrease. The growth curve is more important than the absolute weight.
Feed consumption is often variable due to the fact that it is difficult to measure. Measuring over at least three week gives a good impression. At the start of lay feed consumption should increase towards the laying peak. After this, feed consumption should be stable. During the last quarter of the production period feed is sometimes limited by the farmer. Be concerned when feed consumption decreases.

Analysis criteria for egg production

The production graph is a good aid to farmers as it gives an overall view of the most important technical figures like laying percentage, average egg weight, mortality etc.

This graph should be updated every time the weekly results are calculated. Primary breeders have ready-made graphs with pre-printed standard results. When a farmer plots the results of his flock on the graph he can easily see whether the production of the flock is up to standard or not. But also create your own standards, since the standards provided by the breeding organisations may vary due to type of farm, country, climate, feed quality, etc.

Age?

Age in weeks: often measured from Sunday-Saturday or in Arab countries Friday-Thursday. However, it is best to take day of birth as day 1.

Example of data collection

Age (weeks)	HD prod (%)	Egg wt (g)	Weekly mortality (%)	feed intake (g)	Body wt (g)
18	5	51	0.2	92	1320
19	23	54	0.1	96	
20	48	55	0.3	101	1350
21	64	56.5	0.4	107	
22	86	57.2	0.2	109	1370

Observed problems and possible causes

Delay of the onset of the laying period
- diseases
- hens are not well developed/slow maturing
- low uniformity flock
- poor quality of rearing management
- hours of light are decreasing
- feed utilisation/poor feed quality

High mortality
- poor de-beaking (cannibalism)
- high stocking density
- disease
- housing conditions (draught, light)

High feed consumption
- poor feed quality
- quality of feeding equipment (wastage)
- improper balancing of ration
- improper feed storage
- nutrient deficiencies

Many second grade eggs
- not enough Ca in feed or poor calcium sources
- age of the birds
- high temperatures
- management of litter, nests and birds (not enough nests, poor nest construction)
- diseases

Inspections outside the house

A well-run farm revolves around good inspections. Keep checking that nothing is going wrong or whether there are things that need improvement. Start your inspection outside the poultry house and continue until you are right up close to the birds. Here are some important points to remember.

Rodent problems? Leave an open space of at least two metres around the poultry houses, because mice and rats don't like to cross such large spaces. Also make sure the space around the poultry house is clear and tidy! Or keep any vegetation between houses to a minimum.

Simple solutions

Problems with the temperature of the drinking water? Sometimes simple measures/actions can make the difference, like painting a black reservoir white to reflect the heat from the sun. Check that the water tanks and pipes are properly protected from the sun. Determine the temperature of the water as it enters the house now and again.

At the entrance to the house and the service room:

1. Is the service room tidy?

2. Is the entrance properly disinfected (use a disinfection bath or mat, but make sure you can't walk around it, which is possible in this case...)?

3. Is there a proper system for removing the eggs and dead birds? Do you use new egg trays?

4. Is the key data for checking the results available and entered properly, i.e. over the last four weeks: % mortality, feed intake, production?

Standing or not standing? Do all the birds stand up when you pass by? Weak birds often remain sitting on their hocks. And are the birds curious enough?
Calm or commotion? Walk through many of the aisles, both at the front and the back and observer the reactions.

If eggs are in groups of 3, 4 or even more, the cage surface is forming a funnel which can cause the eggs to roll against each other and break.

Inspection in the poultry house

In cage systems, behaviour is more difficult to monitor, so your senses must be even more alert. Start by checking general aspects..

Eggs

The eggs are the product you are working for. So keep a close eye on them. The eggs will also give you signals that can help you improve your management.

What is the egg shell quality like? Smooth or rough? Misshapen eggs? Are there small holes in the shell? Is there any egg pecking going on? If the shell is weak, check the calcium content of the feed and add seashell grit (or another source of calcium) if necessary, preferably towards the end of the day. Provide a 'midnight snack' to older flocks. This might also help in periods with hot weather when feed intake during day time drops.

Look everywhere

Make sure you can see right into all the cages, not only those at eye-level but also those at the top and bottom. Use appropriate equipment to help if necessary.

Calm or commotion?

Walk through a lot of the aisles, both at the front and the back.

Feed consumption

- Insufficient feed intake: Feed the birds more often during the day and increase the number of hours of light. Also try adjusting the amount of feed.
- Feed intake too high: Do the hens select feed particles? You will see this on the floor. Reduce the feed level in the feeder and also the number of feedings if necessary. Make sure the feed troughs are empty for a while (at least one hour) in the middle of the day.

Do you have weight information: are they growing enough? What is the temperature in the house? This affects the feed intake.

Paying attention to the hen

Before you can judge accurately how a good hen looks like and what behaviour she should be showing, you need some background information. Every chicken has a number of basic needs that can be summarised as feed, water, light, air, rest, space and health. In every housing system and in every situation you need to check these conditions to make sure the chicken's needs are being met. The more free the system, the more attention you will have to pay to the various factors. Because even though a floor system may provide each bird with enough space on average, if chickens are huddling some will suffer. So does every bird have enough space all the time?

These young hens are looking at the entrance of the feed through into the poultry house. They are hungry and it is almost feeding time.

Behavioural needs: preening, dust bathing

Chickens keep their feathers in good condition by preening and dust bathing. Preening transfers the fat from the preen gland to the feathers. Dust cleans the feathers and removes the old fat and any parasites with it. This improves the feather quality: the feathers remain looser and retain warmth better. Dust baths prevent the feathers from becoming brittle, reducing breakage. Chickens start dust bathing from 4-6 days. This can only be done in litter that is sufficiently fine, such as sand or peat. Wood shavings and straw are not suitable. A dust bath also makes financial sense: good plumage is good for the hen and saves you money on feed. The birds cannot dust-bath in conventional cage systems. Special facilities need to be provided, like in furnished cages, which pushes up the cost.

The hen ruffles her feathers to get the litter right inside.

Wish list

To create an optimal environment for your hens, you first need to know their needs. Studies have been performed of the amount of effort they are prepared to make for things. This gives a good picture of their needs, for example by getting them to push a heavy door aside to get something, creep through a narrow opening or getting them to peck at a knob frequently. A hen goes to just as much trouble with its laying nest just before it lays an egg as it does with feeding after fasting for eight hours. This tells us that a laying nest is an important need for a chicken at that time. Based on those studies you can draw up a wish list.

Wish list
1. Laying nest
2. Scratching
3. Perch
4. Dust bath
5. Extra space

Bird behaviour

Cage systems restrict the bird's movement and behaviour. Also this 'boredom' causes undesirable behaviour more likely, like pecking at the vent, claws and eggs if they get the opportunity. It is more important to have a good understanding of your birds' natural behaviour when they are housed in floor systems. Try to recognise undesirable behaviour in time and know what you can do to control it. Even more important is to set up your house and flock management to avoid undesirable behaviour like hens crawling over each other or laying eggs on the floor.

Group behaviour

A chicken is a social animal. She recognises about 80 others of her kind and knows who are the dominant birds in a small group. In cage systems, groups are small but there are no refuges or resting places. In narrow cages in particular, there is a risk that the same chicken will always be stuck at the back of the cage. In that case it is better to put one hen less in the cage. In larger groups, chickens are unable to recognise a clear pecking order among all the individuals, unless there are possibilities for subgrouping in the house. In large groups they tend to form subgroups in which the birds know each other and have a set pecking order. Heavier hens or hens with larger combs have a higher rank. Pecked and weaker hens hide under the slats, in the laying nests or on the racks. Prevent this behaviour by creating additional refuges and resting places, for example on the top tier (with water only) or by installing perches on the slats.

Not all the hens feed at the same time even when the cage is wide enough. Dominant hens feed first. So give them the feed in portions so that even the lower-ranking hens get some.

Behavioural needs: working and feeding

Chickens spend most of their time foraging for food. In the wild, they would spend half their time scratching and foraging. Even though they get their food from a feeder they still love scratching around, including between feeds. In a floor system, make sure they can do so by providing loose, dry litter. Or provide a bale of straw/alfalfa. Then they will also be less inclined to pull out each other's feathers. But make sure the alfalfa is dried and of good quality (no fungal growth).

How chickens spend their time in different environments

Type of chicken	Environment	% foraging for food and feeding	% abnormal behaviour	% resting behaviour
Jungle Fowl (wild chicken)	Wild	50	0	<50
Jungle Fowl	Zoo	60	0	10
Layer	Floor system	18	0	3
Layer	Cage	22	0	8
Chicks of layers that have run wild	Wild	53	0	39

Source: P. Koene, in Bels, 2006

Strict daily routine for layers

Floor housed hens start each day by feeding and inspecting the nest box. Then they lay an egg. In the middle of the day they have a rest and take a dust bath. At the end of the day their scratching and feeding behaviour peaks. If there is any feather pecking, it will usually occur in the afternoon. The afternoon is therefore the most important time to provide distraction. In cage systems there is little or no evidence of a daily behaviour routine. You'll sometimes notice that some hens get a little agitated just before laying because they can't find a nest.

Do your inspections at other times as well, for example after feeding or in the evening when the birds are roosting. Chickens are creatures of habit and do different things throughout the day (see daily routine below). If you do too many inspections at set times you could miss important signals because they are not demonstrating a particular behaviour at that time. For example, you will only observe distress caused by red mites after the chickens have gone to roost. Sometimes you will want the hens to be moving about so that you will notice any dead and passive ones. So make sure to also start going around when the feeding system starts running.

Laying an egg

Laying an egg is quite an exercise for a chicken every day.

1. The hen goes into the nest.
2. She sits there quietly for half an hour or more, often with her eyes closed.
3. She becomes increasingly restless, flicking up her tail repeatedly and spreading out the feathers on the laying stomach.
4. Suddenly she stands up and spreads her legs.
5. She strains at intervals and the egg starts to emerge.
6. The still damp egg pops out, followed by a red mucous membrane.
7. After a few seconds the mucous membrane is retracted and the vent closes.
8. The hen stands up over her egg and rests, beak open and panting.
9. She inspects the egg and leaves the nest, sometimes with a loud cackle.
10. The hen feeds and drinks and resumes her daily routine.

Phases 3 to 6 usually take no more than 30 seconds. Chickens are vulnerable during laying so you should leave them alone.

Daily routine

Night   **Morning**   **Afternoon**   **Evening**   **Night**

Sleeping

Preening

Feeding

Egg laying

Dust bathing

Sunbathing

Mating

Feeding

Sleeping

Anatomy

When you talk to other people, try to use the correct terminology to avoid confusion. To be able to observe irregularities, you first need to be able to recognise what is normal.

1 beak	11 breast
2 nostrils	12 wing
3 comb	13 preen gland
4 ear	14 tail
5 earlobe	15 vent
6 wattles	16 belly
7 hackle	17 shank
8 neck	18 footpad
9 back	19 toe
10 shoulder	

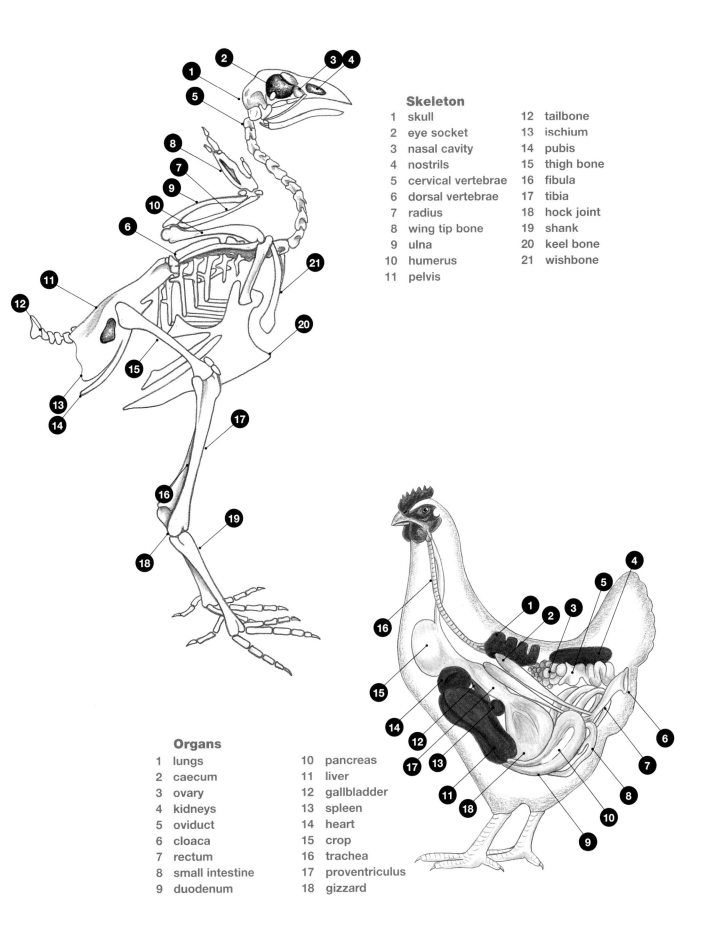

Skeleton

1 skull
2 eye socket
3 nasal cavity
4 nostrils
5 cervical vertebrae
6 dorsal vertebrae
7 radius
8 wing tip bone
9 ulna
10 humerus
11 pelvis
12 tailbone
13 ischium
14 pubis
15 thigh bone
16 fibula
17 tibia
18 hock joint
19 shank
20 keel bone
21 wishbone

Organs

1 lungs
2 caecum
3 ovary
4 kidneys
5 oviduct
6 cloaca
7 rectum
8 small intestine
9 duodenum
10 pancreas
11 liver
12 gallbladder
13 spleen
14 heart
15 crop
16 trachea
17 proventriculus
18 gizzard

Respiratory system

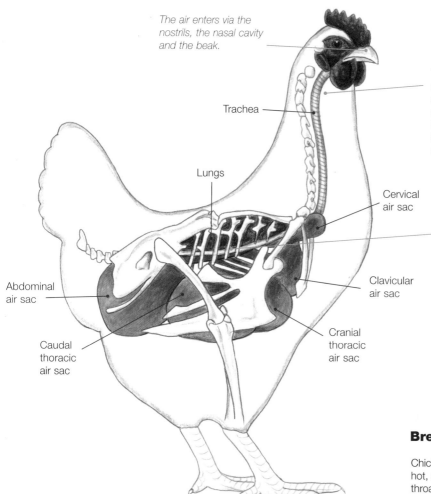

The air enters via the nostrils, the nasal cavity and the beak.

The inhaled air is purified in the trachea which is covered with a layer of mucus and cilia. The importance of the cilia is often underestimated. Disinfecting the hatcher with formalin can affect the cilia. So the use of formalin is not recommended.

Trachea

Lungs

Cervical air sac

A chicken's lungs are relatively small and are virtually unable to expand. Unlike mammals, the lungs of birds end in air sacs. The air sacs are balloon-like features distributed throughout the body. They force the air through the lungs twice, helping it to penetrate deep into the body, which makes the chicken susceptible to respiratory infections.

Clavicular air sac

Abdominal air sac

Cranial thoracic air sac

Caudal thoracic air sac

Breathing to cool down

Chickens cannot sweat. When a chicken gets too hot, she will pant with her beak open (moving her throat backwards and forwards rapidly). This gets rid of excess heat by enabling moisture to evaporate through the airways. At the same time she will often raise her feathers and hold her wings out slightly from her body, allowing her skin maximum contact with the ventilation air to expel heat.

Above 30°C: danger zone

Temperatures of above 30°C in the poultry house are risky, particularly in high relative humidity (RH). Large temperature fluctuations can make it hard for a hen to adapt her body processes quickly enough. However, hens can 'learn' to resist higher temperatures of up to as much as 38°C if the temperature increases gradually.

Senses

Most of a chicken's senses work differently from those of humans. Its eyesight is much better developed, for example, but its hearing possibly less so.

Nosc - smell

Chickens have a good sense of smell but not as good as that of mammals. Chickens use their nose to search for food and to recognise others of their species. They can not only smell high concentrations of substances such as ammonia or carbon dioxide, they also have special nerves which make it painful to do so.

Eyes - sight

Chickens can see many more details and more colours and can see fluorescent lights (105 Hz) flickering, which we humans can't see. Chickens can also see ultraviolet light and are more sensitive to other colours than we are. What we experience as white light can be light blue or red to a hen, depending on the light source.

Tongue - taste

Chickens taste with taste buds. A chicken has 350 at most, a human 9,000. Like humans, chickens can distinguish sweet, salty, sour and bitter.

Beak - touch

Chickens can distinguish several contrasts with their beaks: hard/soft, hot/cold, structural differences (rough/smooth) and pain. The tip is the most sensitive part of the beak. Trimming or clipping beaks therefore causes them pain.

Ears - hearing

Humans can hear slightly higher tones than chickens can. The sounds a chicken makes are between 400 and 6,000 Hz (low tones).

Soil and air-borne vibrations

With the legs and to a lesser extent via the skin, chickens are able to feel vibrations in the ground and in the air. This enables them to detect prowling predators in the dark.

Field of vision

Chickens have panoramic vision of about 300°, but the overlap between the two eyes is minimal. They can only see depth in a narrow angle (shown in green). When you go into a poultry house, you may sometimes see all the chickens briefly shaking their heads at the same time. This probably enables them to see what is happening better (with depth).

Field of vision without depth

Depth perception

Checking individual birds

Make sure you always keep your finger on the pulse, for example by picking up about 20 hens every week and checking them for anything unusual. Take hens from different locations in the poultry house, and if you have cages, from different rows. You don't need to check for everything all the time, but train yourself by checking more and more aspects each time. Make a note of things that catch your eye so that you can check whether they are still the same next time.

The hen should stand tall. When huddled up she is unwell. If a bird stands on one foot for a long time, she could be suffering from stomach pain. Sitting on hocks is a first indication of leg problems (calcium deficiency?).

Catching and holding a chicken

You catch a chicken by approaching her quietly and slowly and quickly taking hold of her by a leg.

It's best to pick a chicken up by the wings as the wings are stronger than the legs, which can break easily. When you pick up a chicken, a healthy one will offer some resistance; you'll feel the strength in the wings.

Red comb:
good.

Pale comb:
possible intestinal dysfunction.

Blue comb:
possible E.coli-infection.

Shrunken comb:
dehydration.

Chicken sounds

Thirty different chicken sounds have been identified. Being able to identify sounds enables you to pick up a number of signals. If your poultry house is peaceful and healthy, you will hear a kind of melodious song coming from it. If you hear the occasional loud screech, be on the look-out for feather pecking. Unusual sounds like sniffling are often a sign of illness.

First impressions

Feathers

Take a good look at the moulting stage of the wing feathers.

Feathering good or bad? How is the feather coverage? Is there any pecking going on? If there are signs of cannibalism, reduce the light intensity if possible.

Tail: well feathered and not pecked.

Legs

Swellings or scabs on the footpads are a sign of wet or sharp litter or sharp protrusions. Smooth, shiny scales are good, but noticeably dry scales are a sign of general dehydration. No sores on the toes or footpads.

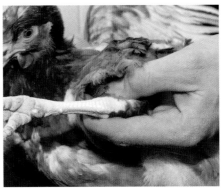

Stiff or hot joints may well be inflamed.

Check the nails; they should not be too long.

Crop

Feel the crop to check whether the hen has eaten enough. If the crop feels too hard the hen isn't drinking enough. It should feel a bit soft.

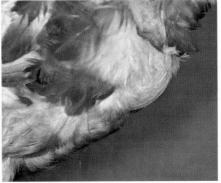

Empty crop

Well filled crop

A closer look

During rearing the breast feels bony, but after rearing the chicken should be fleshy and should start developing a belly.

During laying, a sharply protruding breast-bone and too little fleshiness indicates a protein deficiency. The breastbone must be nice and straight. Breaks are caused by accidents. A soft tip on the breast-bone is either the result of insufficient calcium, phosphorus or vitamin D in the feed or poor uptake from the intestine.

If the space between the leg bones is narrower than two fingers, the chicken is not laying. The cloaca will also be dry. A good layer's leg bones can be moved smoothly and the distance between them is more than two fingers. A good layer will have some fat around the leg bones; if the leg bones feel bony, the bird is too thin.

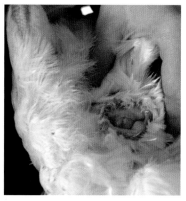

Check that the cloaca is moist, supple and well feathered. No traces of blood from cannibalism or vent rupture.

Push against the beak to check how strong it is. If the beak can be 'pushed over', this indicates a lack of vitamin D3 in the feed.

Comb and lobes. A healthy comb stands upright and has a nice red colour. There should also be a red rim around the eyes. A very large comb can flop over, but that is not abnormal.

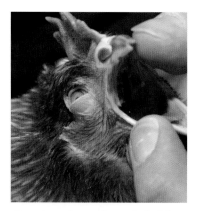

If you hear abnormal sounds, look for wet noses and check the throat for mucus or other signs of inflammation.

Nose and sinuses. A dirty, wet nose and swollen sinuses indicate infections of the airways.

Eyes. Sagging eyelids or moist eyes are an indication of inflammation of the airways. The pupil should be round and clear. If there is matter adhering to the eyes, this is a sign of moist eyes and therefore of eye or respiratory problems.

Weight and uniformity

Weight and in particular fleshiness are good indicators of the hen's condition. You can't see uniformity but you can measure it by weighing birds individually. The less the difference there is between the birds, the easier it is to look after the flock properly. Each bird should really be looked after differently depending on its development. But by ensuring that all birds develop as uniformly as possible, you can treat them more or less as one bird. Uniformity is a criterion for the quality of your management. In other words, whether you are managing to provide each bird the same amount of feed. Check the fleshiness of all your birds. Check the animals and the feed at various locations in the poultry house. Check that there is enough good quality feed everywhere. Weigh the birds regularly to determine how their weight is developing.

Subtle signals of disease

If birds are seriously ill, you'll notice the symptoms immediately. However, there are some diseases such as Infectious Bronchitis (IB) which you can't identify immediately just by looking at the animals. You will observe lower levels of feed and water intake, or you will notice that the shell colour of the eggs has changed or the eggs show defects. If the chickens start laying fewer eggs or laying later in the day, then there has been a problem for a while. This is a sign that there is something wrong and that you have missed something.

The weighing scale in this cage is in the shape of a short perch which the chickens like to sit on.

In cage systems with tiers at different heights, the temperature is higher at the top than at the bottom. The hens in the bottom cages will need a few extra grams of maintenance feed per day. Rule of thumb: each chicken needs one gram of extra food per day for every degree of lower temperature.

The nails of these chickens are too long. Fit new abrasive strips.

Keeping nails trim

In floor systems nails stay short as the hens scratch through the litter and scrabble across the concrete floor. In cages, chickens can only keep their nails short by providing special facilities like abrasive strips on the egg protection panel which the hens scrabble across while eating. Materials used for this include scouring pastes, hard metal strips with a rough surface and stone abrasive strips. Perforated egg protection panels are slightly less effective and are actually only used for brown hens as the nails on brown hens grow more slowly. Adhesive strips are cheap but they generally do not last more than one laying cycle.

Manure

Chickens produce two types of manure: intestinal ('the usual type') and caecal droppings. Intestinal faeces are solid with a white cap of urates. Caecal droppings are more of a shiny, firm paste, dark-green to brown in colour. The manure is not normal if it is milky white, green, yellow or orange or if it contains blood. The manure is also not normal if it is not firm enough, too wet, foamy or poorly digested (feed-coloured or still containing feed components).

Intestinal droppings

Caecal droppings

Scores for intestinal droppings

You can pick up and roll a good dropping in your hand. That does not apply to caecal droppings, of course. If the manure is not in neat pellet form, the chicks are cold, they are sickening or the feed is not optimal. Roughly speaking, the quality of the manure can be assessed as follows:

Right

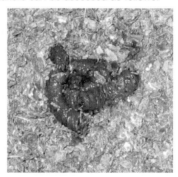

Reasonable

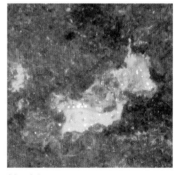

Not right

Source: A. Slaats

Scores for caecal droppings

Caecal droppings should be dark brown and not thin but sticky. If the caecal droppings turn lighter in colour, digestion is not optimal and there are too many nutrients remaining at the end of the small intestine. These will ferment in the caecum and will cause overly thin caecal droppings.

Right

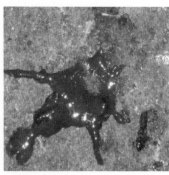

Reasonable

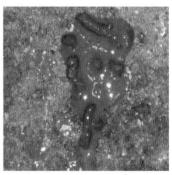

Not right

Source: A. Slaats

Signals from manure

Fresh manure can tell you a lot. A healthy, well-fed and well cared-for chick produces manure in neat droppings. Assess normal (intestinal) droppings and caecal droppings separately.

Manure on a manure belt

You can check the quality of the manure immediately and accurately on a manure belt. So stand next to the belt when it is running and take a close look at the manure.

Manure under the cages

If the manure stays on the floor under the cages, it is harder to check than when it is on a belt. Make sure you look at piled up manure as well.

Manure and litter

Litter should be slightly damp and warm. In places where the litter is wet or that are avoided by birds, there will be no heating effect and the floor will be cold. Birds will avoid those places even more and condensation will cause the litter to deteriorate further.

If you see poor quality litter under the lamps, that means that the light intensity varies too much throughout the house. Birds avoid places with too much light from above, the litter cools down and condensation occurs. Once this starts to happen, the problem gets constantly worse. Remove wet areas of litter. If dimming the light doesn't help, cover over the underside of the lamps to avoid direct light. If you can clearly see entire strips of poor litter, this is often caused by poor air movement. The distribution of the birds therefore has a direct impact on the litter quality, and vice versa.

Wet litter everywhere

If all the litter gets wet, this can have various causes:

- Poor feed composition or demixing in the feeder, making the manure too wet;
- Cold litter, causing condensation;
- Inadequate ventilation;
- Sick birds (producing wet manure);
- Nights too long. The chicks are lying on the litter for too long periods at a time.

Dry or wet manure? Is the manure dry? Is the manure normal?

Wet manure?

Pick up a dropping and squeeze it. In this photo you can see that the pellet contains water. When you squeeze it, it drips: this is not right.

The litter should always be dry and slightly loose. It should not stick to your hands or boots too much. The litter in this photograph is good. From week 3, make a note of the quality of the litter in different parts of the house once a week, marking it on a map of the house, for example.

CHAPTER 2

The hen and her environment

Cage or floor: there is a world of difference between the two. With caged chickens, the birds have no choices and it's mainly the farmer who calls the shots. In floor systems the hens can decide for themselves where to walk, lay their eggs or drops their faeces. As a poultry farmer you must bear this in mind and respond accordingly. You can influence their behaviour with food, water, light and other factors.

Good management requires thinking about things from the hen's point of view and looking out for them properly.

There are also great differences within cage and floor systems. In traditional cages the chicken can't do very much, but in bigger systems containing more than 40-50 hens, you need to take the birds' behaviour into account. In floor systems it makes a big difference whether the birds are free-range or the poultry house is fitted with a type of aviary system. In the latter case in particular, you will definitely need to take the birds' behaviour into account. The hen prefers her living environment to be designed in such a way that there is a separate area for each activity: resting, laying eggs, scratching, eating and drinking, dust bathing. For resting, laying eggs and dust bathing she needs quiet places where she won't be disturbed by other chickens

coming and going. A healthy accommodation naturally also includes the right temperature and the right amount of light, air, food and water.

These perches are at the top of the house where there are no other facilities. The resting chickens are not disturbed there, so they can get real peace and quiet. In cage systems there is very little room for birds to be able to rest undisturbed.

Differences between husbandry systems

There are various types of husbandry systems irrespective of the climatic conditions. In most countries hens are kept in small cages. In countries where these are banned, mini-aviaries or enriched cage systems may be a permitted alternative.
In floor housing chickens have much more room to move and are more free to exercise their natural behaviour. But in doing so they use more energy and therefore eat about 5 grams more feed a day.

There is also a risk of egg loss because hens can lay their eggs outside the nest. Avoiding floor eggs requires a lot of attention and work, particularly in the early morning. In case of floor eggs start walking through the hen house at about the time they lay their eggs, disturbing the hens that try to lay their eggs outside the nests. The choice of housing systems depends not only on local cost levels but also on society's demands in terms of animal welfare (laid down in law). And the choice must of course suit the poultry farmer and his staff.

Cage system

+ Most efficient method of poultry-keeping
- Birds are limited in their natural behaviour (animal welfare)
+ Less labour-intensive
+ Better hygiene (diseases spread more slowly)
+ Climate easier to control

Floor system

- When something goes wrong the consequences are more serious
- Higher management level necessary: behaviour is an extra factor to take into account
+ Birds can display their natural behaviour
+ Better image (meeting demands of society)
- More labour-intensive: a lot of extra work in the poultry house
+/- Ventilation systems work differently in floor systems (fewer chickens so less heat generated, susceptible to weather influences, reduced pressure ventilation not possible in free-range system)

Meeting the birds' behaviour in cages

Enriched caged systems also feature elements that allow birds to demonstrate limited natural behaviour, like mats for their nails or flaps that give them some privacy.

With caged systems there are several options, from very small cages (5 hens per cage) (picture right) to cages containing 30 or more birds (picture top).

Infections spread faster in a house with non-caged chickens than in a caged system because the chickens spread the germs all over the house and come into contact with other chickens' manure.

Open sided or closed poultry houses?

In open sided poultry houses there is often little that can be done about the temperature around the birds. But it is possible to place fans by the birds to cool them down. Curtains are often used to prevent the house from getting too cold at night.

If the temperature gets very high or low, the feed, the amount of feed and the feeding time will need to be adjusted accordingly.

Rearing

In open poultry houses or houses that allow a lot of light, the day length can't be shortened and you have to deal with the natural day length. Getting the birds into lay at the required time may be a problem, particularly during rearing. A modified lighting scheme may help.

In closed houses the climatic aspects can be regulated, so you are in better control of your production. But a reliable power supply is essential and the costs of housing and climate control are high.

In open houses climate control and lighting costs are limited, but so is your control over them. In many countries birds are housed in open poultry houses. The light can't be dimmed.

You can restrict the influx of direct sunlight with a roof that extends well beyond the edge of the building. When building a new poultry house, make sure the length of the house runs east-west. This is another way to restrict light influx at noon.

Climate management in high temperatures

The optimum house temperature for laying hens is about 25°C. Up to a temperature of about 30°C you will notice that the hen can still regulate her own temperature reasonably well. Above that you will need to provide cooling. A desert cooler can lower the indoor temperature quite significantly when the outside air is hot and dry. Make sure the incoming cooled air is not directed straight onto the birds.

If there is no cooling facility, ventilation can help prevent heat stress.

Chickens cool themselves down by evaporating moisture by panting. When chickens pant, the air around the head contains high levels of moisture. It is important to get rid of this air as quickly as possible so that more moist air can be given off. A few extra fans that provide more air movement and are directed at the birds are an effective way of dealing with high temperatures. You should see the feathers moving in the airflow. As they can't sweat, excessive cooling won't affect the birds as long as the air temperature is higher than 26°C.

If it gets warmer or too warm, in floor housing you will notice that the birds are dust bathing more, ruffling their feathers and spreading their wings. Birds in a cage system are not able to do this, but they pant with their beaks open and spread out their wings. It is important to make sure that every bird in every cage has the best possible ventilation.

In intensive sunshine the roof of the poultry house can get very hot and the incoming air around the house can heat up to quite a high level. Spraying the roof creates a lot of condensation which cools it down. Atomisers installed under the roof of the house can also provide some cooling through humidification. But only when the relative humidity is not too high.

These windows can be opened, but they can cause draughts and can only be open or closed. There is little control.

Provision for cool nights

In tropical regions the outside temperature can often drop well below 25°C at night, and even to around freezing point. In open sided poultry houses you will therefore need to close the curtains at night. In this photo you can see that these curtains can be pulled up. The cold air flow should not be directed at the birds. Birds in a strong current of air below 25°C can suffer from problems with colds (mucous) and sometimes coryza. In mechanically ventilated poultry houses, air inlets should be designed in such a way that the incoming air is directed to the ridge of the house where it will be mixed with the warm indoor air.

Ventilation

The house climate is determined by the combination of ventilation, heating and cooling. The choice of the ventilation system should be suited to external conditions. Simple or complex, the system needs to be managed. And even with a fully automatic system your own perception remains crucial (ears, eyes, nose and skin).

Natural ventilation

Natural ventilation doesn't make use of fans for incoming or outgoing air. The fresh air enters the house via open inlets, often fitted with controlled valves, panels or curtains. And leaves through the same openings and/or the roof. Natural ventilation is often seen as a simple and inexpensive system. Whether that is true depends on the results that can be achieved from such a poultry house.

Mechanical ventilation:

Even in areas where natural ventilation can work well, farmers are increasingly using mechanical ventilation. The investment and energy use are higher, but it offers more control options and thus it is more likely to have good results. The air is expelled by ventilators: the key word is underpressure. The slight negative pressure in the house pulls air in via all openings at a similar rate. It is therefore important that there are no openings besides the air inlet valves: they could disrupt the entire system!

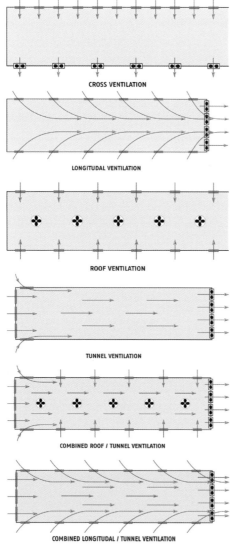

Cross ventilation: The air is expelled on one side of the house and fresh air enters through inlets on the opposite side. This system allows small, but also large quantities of air to move.

Longitudinal ventilation: The air intake valves are on the both sides of the house. The fans are placed in the rear wall. This system does especially well in those areas where the temperature differences are not too large (such as a maritime climate). Investment and operating costs are relatively low.

Roof ventilation: The fans are installed in ventilation ducts in the roof. The air intake valves are evenly distributed over the two sides of the house. This is often used for minimum ventilation in colder climates. Small quantities of air can be well managed. For larger quantities of air, the system is often more expensive because of the large number of fans and air ducts that are required.

Tunnel ventilation: The fans are placed in the back wall and air is sucked in through air inlets in the front wall of the house (or in the final few metres of the side walls at the front end). This creates relatively high air velocity. This high air velocity (up to 3.4 m/s) gives a cooling effect on the animals (chilling effect). This system is used when large quantities of air are required.

Combinations: Tunnel ventilation is often used in combination with a roof or longitudinal ventilation, for example. In that case, the roof/ longitudinal ventilation is used for minimum ventilation. When more is required these valves are closed and the tunnel inlets opened. An increasingly used concept.

Assessing ventilation in the poultry house

In floor housing systems, the distribution of birds over the house tells you something about the quality of ventilation. But you can also assess the ventilation in other ways.

Enter the house with bare, wet arms or wearing shorts, go and stand in the parts of the house where there are too few chicks and feel whether there is a draught there. Check whether the litter feels cold. See whether there is a pattern in the way the birds are distributed throughout the house and whether this has anything to do with the position of the lamps, fans, air inlet etc. If you change the settings, give the chicks a couple of hours to adjust. Don't conclude too quickly that the change has not worked or is no good. Make a note of what you have changed.

Avoid ventilation errors. Check all your equipment at set times.

Good ventilation

Poor ventilation with floor rearing

If you cannot feel it yourself, do a smoke test to see how quickly the air is flowing through the house. You don't have to take the chicks out of the house to do this. There are several options:

- The fresh, cold air in the middle sinks and there is little air movement at the sides.
- The chicks avoid the middle and go to the sides of the house, resulting in damp litter.
- Reduce the underpressure.

- The fresh, cold air sinks too quickly and is therefore not being heated up enough. The chicks keep to the outermost quarters and the middle of the house.
- This has created two empty strips down the length of the house: a zebra crossing effect.
- Increase the underpressure.

- The chicks move away from the edges and are mainly in the middle.
- The flaps are too tightly shut so there is too little air entering through each flap which dissipates too quickly.
- Open some of the flaps about two fingers more.

- In hot weather the flaps will turn.
- The air will pass right over the chicks at high velocity.
- This will make the air feel quite cool near the birds (wind chill effect). This should only be done deliberately if the ambient temperature is very high.

Source: Henk Rodenboog

Climate under control

The climate in a house is a combination of temperature, air velocity, indoor air composition, dust and light. And very important: the (micro) climate around the birds is what counts! These factors can impact on one another. Get a climate expert to check both the climate computer and the climate once or preferably twice per year. The expert works with these systems every day and knows what the best setting should be. Sometimes it will be different from the manufacturer's recommended setting. The expert can also pick up on changes in the sensors which could indicate that the climate is no longer being optimally controlled. Naturally you should also be alert to signals that indicate whether or not the climate is right. Chickens might avoid certain places or huddle together, for example, or there may be a stuffy smell. You get used to bad air quite quickly, so go with the impression you get when you enter and leave the house.

When setting the house temperature, take the quality of the plumage into account. A featherless hen needs a higher temperature.

When setting the house temperature, take the quality of the feather cover into account. A featherless hen needs a higher temperature and is more susceptible to draughts and air flows.

Air flows

Wrong

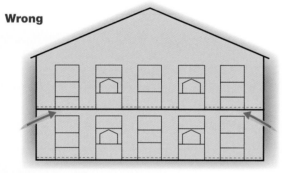

Bad air circulation can occur in houses with little volume and a relatively large number of obstacles. Air does not circulate properly in aviary houses that are too low. There are also places with no air movement in the middle of the poultry house.

Right

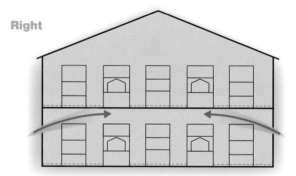

There is plenty of room above the tiers to allow the air right into the middle of the house. It is also less likely that there will be places without air movement. You can also direct air to the middle of the house with pipes or ducts from outside running along the ceiling to the middle.

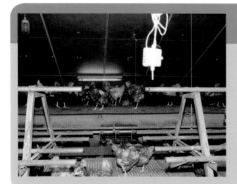

LOOK-THINK-ACT

What is wrong with this temperature sensor?

This temperature sensor is level with the top perch. That's too high. For accurate temperature measurement, it is important that the temperature sensors register the temperature where the chickens are. So it must be in among the chickens, but not in a position where the chickens can sit against it. Check regularly that the temperature sensors are working properly by hanging a good manual thermometer next to them.

Effective temperature (wind chill factor)

As it becomes warmer, poor ventilation can cause the air to become musty. This is one of the causes of feather pecking. So you will need to ventilate the house well, ensure a good air velocity and monitor the temperature. In a closed house, make sure the set minimum ventilation is appropriate to the number of hens, and assume 0.7 m³/kg live weight per hour. Ventilation-directed air flow has a cooling effect on the hens, because the wind chill factor increases as the air velocity increases. Watch out for draughts. In houses with floor systems hens will avoid draughty places. The optimum effective temperature for hens in cages is 20 to 24°C. For hens in floor systems it is 18 to 22°C. Higher temperatures over long periods of time, particularly above 28 to 30°C, combined with high relative humidity can result in heat stress. In case of acute heat stress, hens sit with their beaks open and their wings spread out. Mortality is increased and production drops. Chronic heat stress has more gradual effects.

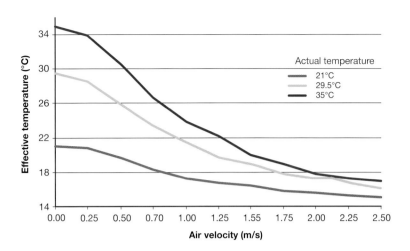

The temperature the chickens experience depends on the combination of outside temperature, relative humidity and air velocity. Higher air velocities in high outside temperatures can create a strong cooling effect. But watch out for draughts.

Wind: not too little, not too much

One disadvantage of natural ventilation is that there is virtually no ventilation when there is no wind. Use auxiliary fans to ensure sufficient air circulation. Fresh air can also reach the birds via the aeration of the manure belts.

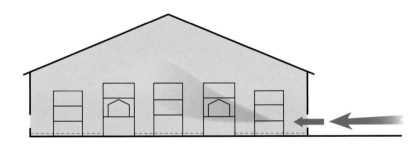

In houses with natural ventilation, the wind affects the interior climate. Too high air velocities can create draughts, and draughts can also pop up at different places in the house.

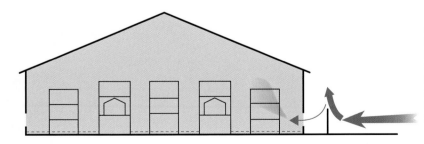

An evergreen windbreak or vertical walls outside the range openings around the house is one way of reducing the effects of wind. In closed poultry houses, wind protection can be fitted in front of the air inlets.

Air

High ammonia and dust levels affect the chickens' mucous membranes and make them more susceptible to disease. In fact, too much ammonia can even cause blindness in chickens. You can smell ammonia from a concentration of as little as 20 ppm yourself. If you can smell it, the concentration is too high. Other gases like oxygen, carbon dioxide and carbon monoxide have no smell. Imperceptible to humans, excessive concentrations can be harmful to the birds, but also to humans. Ventilation not only brings fresh air into the house, it also removes residues in the air.

If you have several poultry houses, you may sometimes notice the hens in different houses doing different things. If the birds in one house are less active than those in another, this may well have something to do with the climate. Monitor it, or get someone in to do so, and improve your ventilation system if necessary. You can monitor most gases easily yourself using gas detection tubes. The table gives the standard levels for the main gases.

Avoid wet litter

Wet litter is a source of ammonia and leads more quickly to digestive problems, coccidiosis and footpad problems (lameness). Keep the litter dryer by ventilating better (removing the moisture), increasing the fibre content in the feed (result: dryer manure) and stopping spillages of drinking water. Scatter grain so that the hens scratch the litter loose themselves.

Concentrations of various gases

Gas	Standard level
Oxygen (O_2)	> 21%
Carbon dioxide (CO_2)	< 0.2% (2000 ppm)
Carbon monoxide (CO)	< 0.01% (100 ppm) (ideally 0)
Ammonia (NH_3)	< 0.002% (20 ppm)
Hydrogen sulphide (H_2S)	< 0.002% (20 ppm)
Relative humidity	60-70%

You can see that the litter in this picture has become a hard crust. The litter was originally damp and was obviously not worked enough by the chickens.

When you monitor the house climate, don't only do so at your own working height but also at the height of the chickens. With cage systems, take measurements at the bottom layer and the top layer.

Light

A properly lit house gives you a good overall view of the house and your hens. This applies to all types of housing systems. With floor systems, by distributing light throughout the living space and over time you can influence where the birds go and when, and whether they are active or rest. Day length and light intensity influence feed consumption and production.

Did you know that a hen...

- sees more colours than humans in daylight?
- sees the light of conventional fluorescent tubes as flickering? This does not apply to high-frequency lamps, which are also more economical to run.
- prefers at least 60 lux for eating, drinking and scratching?
- prefers semi-darkness for egg laying and roosting: 0.5 to 1 lux?
- recognises others of their species better in more intensive light (> 70 lux)?

Pros and cons of lighting systems

	Incandescent lamp	High-frequency fluorescent tube	Orion PL lamp*	SL-lamp	High-pressure sodium vapour lamp
Purchase price	+	--	-	+/-	+/-
Installation costs	-	+	-	-	++
Peripherals	+/-	+/-	+/-	+/-	+/-
Maintenance costs	+/-	+	+	+	+
Power consumption	--	++	++	++	++
Service life	--	+	+	+	++
Dimming	++	++	+	-	+/-
Light distribution	++	+	++	+	-
Spectrum display	+	+/-	--	-	--
Stroboscopic effect	++	++	++	--	+
Feather pecking/ cannibalism	+	+	++	--	+/-

++ = very good; + = good; +/- = average; - = poor; -- = very poor
* In laying hen houses with the Orion PL system, one half consists of red lamps and the other half of white ones.

LOOK-THINK-ACT

Dark spots?

There are dark spots in some places under the racks in this aviary house. The hens will want to lay their eggs there. If you want to avoid floor eggs there, fit rope lights, for example, as shown in this picture.

Dust

There's no such thing as a dust-free poultry house. Litter, manure, feed and feathers all turn into dust eventually. Dust is bad for the health of humans and hens. Dust particles get into the lungs. In combination with ammonia, which affects the mucous membranes, this increases the risk of infection in the birds. Breathing in dust is also dangerous to human health, particularly:

- in high concentrations
- when you stay in the house for a long time
- with very fine particles.

What starts out as a seemingly harmless symptom like a tickle at the back of your throat, sneezing and coughing can turn into serious illnesses like bronchitis, shortness of breath, asthma or reduced lung capacity. Never underestimate the health risks of dust; it is best to wear a dust mask.

Types of dust

The smaller the dust particles, the deeper they penetrate into the lungs and the more harmful they are.
They are classified as follows:

- Inhalable dust: particles of less than 50-100 µm. You can inhale these particles, but you can also expel them via the cilia in the lungs.
- Thoracic dust or particulate: particles of less than 10 µm.
- Respirable dust: particles less than 4 µm. These are very small particles that come to rest in the alveoli and damage the lung function.

Dirt and dust in inlet valves and ventilation ducts cause more resistance. This reduces the ventilation capacity, so the temperature rises. The electricity consumption will rise unnecessarily.

A lot of dust is released during vaccination. When doing these kinds of tasks, always wear a dust mask. Correct use of a dust mask reduces the risk of inhaling dust by 90%. Masks with an exhalation valve are more comfortable to wear. Use P2 masks as a very minimum.

Keeping the aisles in a house with cages clean helps to cut down on dust. Cleaning weekly prevents large amounts of dust from settling which can be disturbed again. Tip: Use a clean filter. An ineffective or old filter will cause the vacuum cleaner to re-expel some of the dust particles. So clean and replace the filter regularly. And don't forget to wear a dust mask yourself.

Solutions for the future

1. Applying an oil film in floor housing systems: binding dust particles by applying a film of rapeseed oil or sunflower oil over the litter. This reduces the amount of dust by 50-90%. Downside: dirt becomes caked.

2. Water spray: settle dust by spraying with clean water. This reduces the amount of dust by 80% (coarse dust) and 50% (fine dust). Downside: relative humidity in the house can become too high, so the litter can get too wet.

3. Air recirculation with cleaning: outgoing air can be recirculated after cleaning (e.g. filtration, air washing). This reduces the amount of dust by 40-60%.

4. Ionisation: settle the dust by charging the dust particles by applying a voltage difference. The charged particles will then stick to earthed surfaces like the floor and walls. This reduces the amount of dust by about 35%.

These techniques have an additional environmental benefit: much less dust is emitted into the open air.

Spraying the litter with oil

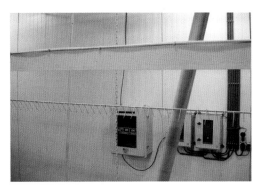

There is a voltage difference along the wire with protrusions (ionisation).

Activities and their dust scores

Presence between hens	Dust score
Delivery of layers	12
Removing birds (catching and loading)	12
Handling individual birds (healthcare etc.)	12
Collecting floor eggs in aviary and floor systems	12
Treating groups of hens (group vaccination)	6
Inspections among birds	3

Other activities in the house	
Cleaning the house	12
Removing dust from aisles, dry	12
Removing dust from aisles, wet	8
Scattering straw and litter	6
Mucking out chicken house with shovel/loader	6
Collecting eggs and inspecting in lobby	4
Inspecting from feed aisle during feeding in cage systems	3
Inspecting from feed aisle outside feeding times in cage systems	2

www.pakstofaan.nl

The dust burden in the house varies depending on the activity. The dust score ranges from 1 to 18, with 18 indicating a very high dust burden and 1 a very low one. These figures do not represent the absolute quantities of dust but a combination of quantity and time. For example, removing hens generates a lot of dust but only for a short period; floor eggs generate very little dust but over a longer period. So the total dust burden of these two activities is the same.

Why free-range?

In various countries society at large has expressed the desire to have free ranging chickens. Consumers want chickens to be able to move about in the open air and not be cooped up in a poultry house all day.

An outdoor range can make a significant contribution to the well-being of the hens. Chickens that get outdoors are less likely to feather-peck. But you need to make the range attractive and safe for the birds and keep it properly maintained.

Health risk

In uncovered outdoor ranges birds are at much greater risk of being infected with bird flu by wild birds. With indoor housing you must also ensure that no other birds can get in and that your birds do not come into contact with other birds through the mesh.

Gimme shelter

Chickens only feel safe when there is shelter nearby. This can consist of natural vegetation or an artificial shelter. What is important is that the hens can stand under or next to it. This makes them feel safe from predators. Then if they get frightened by something they don't have to run indoors. Chickens that only feel safe indoors will hardly ever go outdoors.

In a system with a well-designed outdoor range, hens can exercise their natural behaviour to the full.

There is a gutter under the slats. The chickens cross the slats before they go into the house, so they carry less mud inside on their feet. Discharge the gutter into the manure tank.

Artificial shelter

Camouflage nets (left) disintegrate if they are left out in the open the whole year round, but are a good temporary solution after harvesting or when other forms of shelter are not yet fully operational.

The shelter on this farm (right) is portable. The chickens use it as a refuge when they take fright.

Covered range or winter garden

It's often not possible or feasible to provide a real outdoor range. In that case, consider providing a covered range, also known as a 'winter garden' or cold scratching area. The benefits for the birds are that there is daylight, a different temperature zone and some diversion. You can also use the winter garden to provide the chickens with a diversion:

- bales of alfalfa hay
- freshly mown grass
- barrels of normal grit or gizzard grit
- perches
- containers of sand.

Right: On this farm, a strip of grass is mown every day and fed to the chickens.

Wrong: There is a lot of daylight in this covered range, which makes the chickens active. But there isn't even any litter to scratch in, so the chickens start feather pecking out of boredom.
Inset: The same farm a few months later: lots of bald hens.

Right: In this covered range, trees have been planted which will provide shade in the future. Drinking water is also available.

New housing systems are increasingly designed around the hens' natural needs with a covered range. A covered range doesn't get muddy and prevents infection by wild birds.

CHAPTER 3

Rearing hens

The rearing of pullets is geared towards supplying hens that are healthy and problem-free and that will supply the poultry farmer with lots of good quality eggs.

The circumstances during rearing account for 60-70 per cent of the birds' technical performance later on at the laying farm.

The 16 week rearing period is only a fifth, or even less, of the hen's total life. But it is the most important part. Any mistakes made during rearing cannot be rectified during the laying period and impact severely on the results in the production period. Good preparation starts with drawing a good rearing programme even before the chicks are placed in the rearing house. The very first question is: Do you want an early onset of production with more eggs but a lower egg weight, or a later start with slightly heavier eggs? Controlling growth is an important way of achieving your goal.

Don't forget to take the season into account as well. Birds reared when day length is increasing (spring), even in dark poultry houses, will start laying earlier than those reared in the autumn when the days are getting shorter.

By the time they reach the rearing farm, the chickens will already have gone through quite a lot: all the handling at the hatchery, including sexing, vaccination and possibly beak trimming, followed by transportation, a new environment and changing climatic conditions.

A good start is half the battle

The first five weeks are extremely important for success later on. A hen is kept for up to 95 weeks. Whereas the formative time for a human is the first five years, for a chicken it is the first five weeks. Make a checklist which you can adapt and improve each time.

A flock that gets off to a good start will result in a lower mortality rate and fewer poorly developed birds. Avoid all possible stress during the first five weeks of rearing. This is the time when the most important organs like the heart, lungs, kidneys etc. are developing. Stress slows development and has a negative effect on the laying period.

Three periods

You can divide the rearing period into three periods. During this time the young hen doubles her body weight five times.

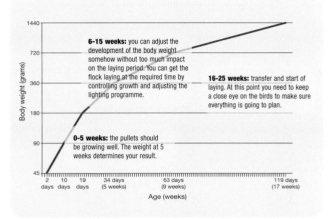

6-15 weeks: you can adjust the development of the body weight somehow without too much impact on the laying period. You can get the flock laying at the required time by controlling growth and adjusting the lighting programme.

16-25 weeks: transfer and start of laying. At this point you need to keep a close eye on the birds to make sure everything is going to plan.

0-5 weeks: the pullets should be growing well. The weight at 5 weeks determines your result.

Brooding rings are used in houses with local heating (often in open houses). Whole house brooding (space heating) can be applied in solid sidewall houses

Rearing in cages or on the floor/litter?

The system you use for rearing your hens has an impact on the results in the laying period.

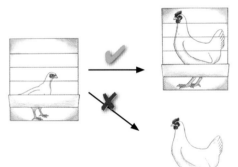

If possible, use the same drinking system in the rearing period and in the laying period. If you change the drinking system, watch out for dehydrated hens.

Hens reared in cages will have problems finding feed and water in a floor housing system later on. There is also a greater risk of floor eggs because the birds have not learned to jump. They will also not have built up enough immunity to diseases like coccidiosis.

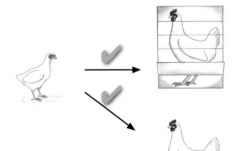

Delivers the best results.

Necessary for successful production in a conventional floor or aviary housing system.

The early days

In the hatchery the chicks are given a range of treatments, so they are very tired when they arrive at the rearing farm and want to rest. Remove the chicks from the boxes as quickly as possible. Put the boxes that cannot be handled immediately in a separate room at a temperature of 22-23°C, not in the poultry house if it is much warmer (above 30°C).

Before the chicks arrive, check that everything in the house is working properly: heating, thermostats, ventilation, feed and water system (water pressure on the nipples, no residues or disinfectants in the water) and lighting. It is also important to make sure the water is microbiologically clean. Because of the high temperature in the house, you will need to change the water one day before the chicks arrive.

Nice and warm

Make sure that the house is nice and warm before the chicks arrive. It is not just the air that needs to be at the right temperature, but the floor in particular and the inventory: the slats, the paper, the feeding system and the drinking water. Cold water (< 20°C) leads to a lower body temperature, which day-old chicks can't adjust. Make sure the air temperature at the level of the chicks is between 33 and 35°C.

In cages the chicks can't seek out a warm place for themselves. So make sure the temperature in every cage or brooding ring is right and that the chicks stay warm. Cold chicks do not eat and drink enough.

Parent flock and chick size

The chicks of young breeder hens are smaller and need a higher temperature and humidity. If you know that, you can prepare the poultry house and the care accordingly to prevent problems.

Yellow chicks are often thought to be more vital than white ones. But this is due to formalin in the hatcher baskets, which colours the down yellow. The colour itself does not make the chicks more vigorous. However, there is an indirect connection: strong chicks hatch earlier and are therefore in contact with formalin for longer and will be a darker yellow colour.

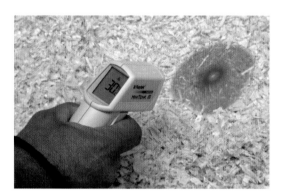

In a floor system, the temperature of the floor should be about 30°C. You can measure this with an infrared thermometer.

These five-day-old chicks are all sticking their heads out of the cage. This could be because they are too hot or the carbon dioxide level in the air is too high.

This method of transportation won't have a happy ending. Chicks will suffocate, or at the very least they will be severely stressed.

Quality of day-old chicks

Assessing day-old chicks individually: what to look for

Check	Right	Wrong
Reflex	Lay chick on its back. It should stand up within 3 seconds.	Chick takes more than 3 seconds to stand up: chick is listless
Eyes	Clean, open and shiny	Closed, dull
Navel	Navel should be closed and clean	Bumpy: remnants of yolk; open navel; feathers smeared with albumen
Feet	Feed should be a normal colour and not swollen	Red hocks, swollen hocks, malformations, deformed toes
Beak	Beak clean with closed nostrils	Red beak; dirty nostrils; malformations
Yolk sac	Stomach soft and malleable	Stomach hard and skin taut
Down	Should be dry and shiny	Down wet and tacky
Uniformity	All chicks the same size	More than 20% of chicks are 20% heavier or lighter than average
Temperature	Should be between 40 and 40.8°C.	Above 41.1°C: too high, below 38°C: too low. Should be 40° 2-3 hours after arrival.

A fit and healthy day-old chick gets on her feet again within three seconds, even when you put her on her back.

Deformed toes are a regular occurrence. The cause can be genetic, but this may also be caused by a vitamin B deficiency or the hatching temperature being too high.

Navels

Check whether there are any chicks with poorly closed navels, for example due to the yolk sac not being fully withdrawn. These navels often do not close at all and pose a greater risk of infection and mortality. So you don't want chicks with poorly closed navels. Make a note of how many there are and discuss this with the hatchery. An open navel with no obstructions will close properly.

When the chicks are hatched at too high a temperature, you will notice red hocks. The red beak is also a signal of this. The red beak happens because the chick wants to get out and tries to stick her head out through the gaps in the crate.

Not good but acceptable: navel will close.

Unacceptable: cannot close because there is some yolk is in the way.

The chicks' body temperature drops between the time when they are removed from the hatcher and when they are placed into the house. Generally speaking, 10% of chicks that come off the truck with a body temperature of 37°C will die. If you are not sure about the body temperature, measure it at the vent with a child ear thermometer.

On arrival, feel whether the chicks are hot or cold. Feel the feet and put them against your lips: this is the best way to feel the temperature. Repeat this when the chickens have been in the house for a couple of hours. If they are still cold, increase the temperature by a few degrees.

Comfort signals in day-old chicks

After being released, the chicks have to get used to their new environment and rest a while before they go exploring. So it is normal for them to sit still for a while to begin with, but after about four hours they should start spreading out, exploring and feeding. If they do not, the litter or the air in the house is too cold. Cold will get them off to a bad start.

If they are sitting too close together on the first day, for example because the temperature is too low, they will go on doing that if you don't react. This sows the seeds for poorly developed birds and therefore an uneven flock. Sitting close together can also cause them to get too hot. Try to spread out the chicks as soon as possible after their arrival by increasing the temperature and dimming the light slightly. If young chicks all press up against the wall, it is too light or it is too warm in the middle of the house. If the chicks are too cold, they will cheep loudly.

Spreading out in the first hours

Four to six hours after being released, the chicks will start to disperse.

The chicks in this house have spread out well in the 24 hours since they were released.

LOOK-THINK-ACT

Why are these chicks all bunched up together?

These chicks are all standing on the last piece of poultry paper. They obviously prefer standing on the paper than on the slats. When rearing in cages, put the chicks at eye-level and cover the floor with paper. A common method is to use several layers of paper and remove the top layer every day.

Temperature

How much you ventilate not only depends on the temperature but also the humidity in the house, the speed of the air flow around the animals and the carbon dioxide level. If the carbon dioxide level is too high, the pullets will become lethargic. If you get a headache after working just above the height of the chicks for five minutes, the carbon dioxide level is at least 3500 ppm, which means that ventilation is poor.

> **Tip**
>
> Get the chicks to spread out more in the house by walking from front to back and tapping on the wall. The chicks will be attracted by the sound and will spread out better.

Ensure adequate relative humidity (min. 55%). In cold periods when you need extra heat you can install a spray head on your heat cannon if necessary, or throw a couple of buckets of water over the aisles and scratching area; that will work wonders.

Young chicks need warmth; they can only regulate their body temperature themselves after 3 to 4 days. Small one-day old chicks, usually from young breeder hens, will need a temperature that is 1-2°C warmer than heavier chicks from older breeder hens. Ask what age the parent flock is. Weigh your chicks on arrival so that you know what you are getting. Light weight chicks need more heat than heavier chicks.

> **Behaviour of a flock of day-old chicks**
>
> Behaviour is a key indicator. Check the chicks' behaviour every couple of hours, not only during the day but at night as well.
> - Chicks are spreading out all over the space: temperature and ventilation are fine.
> - Chicks are huddling together in some places, are less active, don't start moving around and look as if they are in a daze: temperature too low.
> - Chicks are avoiding certain places: it could be draughty there.
> - Chicks are lying on the ground with wings spread out, seem to be gasping for air and start chirping: it is too hot or there is too much carbon dioxide in the air (measure=know).

> **High humidity: higher effective temperature**
>
> The effective temperature (wind chill factor) is related to the relative humidity. How high the relative humidity can be depends on the temperature. The norm for effective temperature is 90 + age of the chicks in weeks. The optimum level is RH + temperature. An example: the chicks are three weeks old, the RH is 73% and the temperature is 24°C. The norm is 90 + 3 = 93. The level is 73 + 24 = 97. Conclusion: the air is too moist, so you should ventilate more.

LOOK-THINK-ACT

House floor wet or dry?

What do you notice about the colour of the house floor? It's dark, so it's wet. That's because the house is too damp. In this case you should ventilate more. Check whether this is happening all over the house or only in certain places.

Weak chicks

Most losses usually occur in the first seven days. If there is a problem with the quality of the parent flock or the hatching conditions, mortality rates can mount up. Good care is all the more important for weak chicks: enough feed and water that they can access easily.

Receive them in a house that has been warmed up well in advance and that has a well insulated floor (chick paper or thick litter) with the temperature on the high side. Young chicks cannot regulate their own body temperature, and if they are not yet eating they will get cold and die.

Signal	Possible cause
Poorly developed birds	Difficulty finding feed and/or water. The feed was not easy enough to reach or the feed on the paper was eaten up too quickly. This kind of problem cannot be fixed in this cycle.
Wrynecks and stargazers	Inflammation of the brain. This can be caused by an infection with salmonella, streptococ, entero-cocs or fungi (*Aspergillus fumigatus*). This is often a hatchery infection.
Lame chicks	Bacterial infection with salmonella, streptococ, enterococ or *E. coli*. A bacterial infection at this age is often related to the quality of the hatching eggs or conditions in the hatchery. Thereafter, the quality of care largely determines the severity of the problem.
Huddled up, feathers raised	The chicks are cold.

Torticollis (wryneck) resulting from meningitis.

Stargazer, also resulting from meningitis.

A bird that has choked and died possibly due to a viral infection or a heart problem.

Vent pasting

A light-grey lump of 'cement', often caused by a serious bacterial infection (e.g. Salmonella) or a kidney disorder. You should rather remove these chicks. An inflammation of the peritoneum affects intestinal peristalsis, which causes the urine to run spontaneously out of the vent. Once dried, it forms a cement-like coating. Often the result of a period of stress.

A chick with a dark-grey 'pencil' formation is not so badly affected.

Cage rearing

When rearing in cages, pay special attention to draughts and lighting. This can be a problem near the ventilation ducts, particularly in dark poultry houses when the pullets search the light of the ventilator. Keep the temperature at chick height at 36-37°C for a few hours. This is higher than in a floor system, because the chicks are not sitting on a warm floor. After this you can follow the general temperature schedule. Cover the floor of the cage with a plastic mat and a few layers of paper, and remove a layer of paper when necessary (every second day). For the first couple of days, spread a little feed on the paper in addition to the feed in the feeding troughs, so that they can easily find it. The chicks should ideally be able to drink from open water.

Space to grow

The living space for chicks is very important for ensuring good uniformity. 125 cm^2 per bird (80/m^2) is enough up to three weeks, but afterwards you should move them to multiple tiers so that they have as at least 220 cm^2 of space (44/m^2). Don't wait any longer than this. The more space the chicks have, the better their weight and uniformity will be at five weeks. Check the equipment of the other tiered cages (drinking nipple/trough, feeding trough height). In cages too, it is important that all the chicks can feed and get water easily at the same time.

Paper in the wire cage provides warmth and protection from draughts (left). It is also better for the chicks' fragile feet. Chicks' feet can get through the mesh (right) and get hurt.

Make sure the air inlet openings are properly closed during the first few days, or there could be a cold draught. Also, incoming light can make the light intensity much higher and the day length longer than it should be. This can cause reduced production when the hens are moved to the laying house, if they arrive in a place where the light intensity is much lower and the day length is shorter.

Put all chicks on one tier during the first two to three weeks, at eye level, and ensure that the lights are hanging low enough so that the chicks have light throughout the whole cage, like here.

Floor rearing

For each heat source serving 500 chicks, fit barriers in a circle approximately 6 metres in diameter. Enlarge this brooding ring regularly after a week, and remove them after three weeks, so that the chicks have enough space.

Make sure the floor is nice and warm and the litter is a few centimetres thick. Hang the heat source roughly at knee height so that all the chicks can sit in the warmth radiated from the heat source. As the chicks get older they need more space, so the heat source should be gradually moved higher up. This increases the size of the floor area heated by the heat source while still enabling all chicks to sit in the warmth at their own optimal temperature. An additional benefit of positioning the heat source higher up is that the temperature under the source on the ground gradually gets lower and you can slowly switch to space heating.

A clear case of draughts. But the corner of this area has been rounded so that the birds cannot be pushed into a corner.

Chick distribution within brooding rings

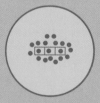

Good: *constantly cheeping chicks evenly spread with a small empty circle in the middle directly below the radiation source. If a lack of space means that there are chicks outside the heat source, these chicks can't sit in the warm area. In that case hang the brooder higher.*

Too draughty: *noisy chicks huddled together away from draught. If chicks are crawling over each other because there is a draught, the temperature in the group will be very high because of a lack of ventilation. This causes the chicks to overheat, which is very stressful for them.*

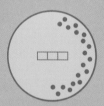

Too hot: *drowsy chicks spread around perimeter. Hang the brooder higher.*

Too cold: *noisy chicks huddled under brooder Hang the brooder lower.*

Signals from the daily mortality pattern

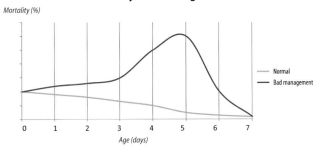

Increased mortality: bad management

Mortality (%)

Age (days)

Normal
Bad management

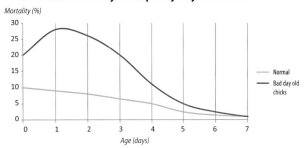

Increased mortality: bad quality day-old chicks

Mortality (%)

30
25
20
15
10
5
0

0 1 2 3 4 5 6 7

Age (days)

Normal
Bad day old chicks

The mortality pattern in the first few weeks gives you a clear signal about your management. Losses in the first three days are closely connected with the quality of the day-old chicks. After three days the losses depend on the quality of care after delivery. Chicks with dirty feathers around the vent (dirty lumps) indicate that they have suffered a period of stress. This problem can no longer be rectified in this cycle. Try to limit the damage as much as possible and make improvements in the next cycle.

From 6-15 weeks

Preparing for the laying period starts with the question: whether to start hens producing early or late? Starting laying earlier gives more eggs in total but with a lower average egg weight.

Postponing start of production will result in heavier eggs, particularly in the early stages of egg-laying.

You can get your hens producing at the required time by controlling growth and adjusting the lighting programme. The age at which you start stimulating egg-laying - by increasing the day length (mimicking the springtime) - depends on the weight of the hen. The optimum weight for this differs between breeds and is specified by every breeder in their management instructions.

Postponing the start of egg-laying

If you want heavier eggs or if you are moving birds into the laying house very late, you can deliberately postpone laying by adjusting the way you manage the birds.

The following factors can have the unintentional and possibly undesired effect of postponing the start of production:

- very hot weather, which can retard growth
- stress caused by coccidiosis or other parasite infections
- late de-beaking due to recurring pecking problems
- severe reaction to vaccination
- problems with the feed

Coming into lay too early is mainly caused by the lighting programme and fast growth.

The more uniform the better

Uniformity is never good enough. Poor uniformity always causes disappointment.

If a uniform flock is lighter or heavier than average, you can easily change that by adjusting how you manage the lighting and feed. But this can't be done if the birds differ greatly in weight. Your uniformity should be at least 80%, but the target is 90%!

Gizzard grit is the chicken's teeth. In cage systems it is difficult to administer gizzard grit if you are using a chain feeder because of the wear on the feeding system. Gizzard grit is not always given as standard in floor rearing systems, even though it can contribute to good digestion. Five grams per bird in the litter at 5-6 weeks and again at 10-11 weeks is plenty.

Weight and condition

When underweight birds come into lay, this can affect their performance because they cannot yet eat enough food. Birds that are too heavy and have more developed laying organs but are not stimulated to lay will eat a lot and will become too fat. Be aware that body weight standards in the rearing period are minimums and maximums for individual hens. If the average weight is around the minimum norm, wait a few days before providing more light, because you will still have too many low body weight individuals. In a cage system, take particular care that the hens in all tiers are the same weight and a good weight.

Causes of reduced uniformity in cage rearing:

- Uneven feed distribution, e.g. caused by very long chain feeders
- Overcrowding. So with cage rearing, always use all the tiers after three weeks.
- Difference in feed structure. You'll only see very fine feed left at the end of the chain feeder and coarse particles at the start.
- Uneven occupation in the various sections/batteries caused by different mortality rates in each cage, for example.

Causes of lower uniformity in floor rearing:

- Overcrowding
- Insufficient feeder space

By the end of the rearing period, the combs are larger and redder. There is also more colour round the eyes, as you can see in this photograph.

Hens moult their wing feathers from the inside out. This hen still has three feathers to moult, namely the long ones on the far right. Long, often pointed feathers are the oldest and have therefore not yet been shed.

Determining uniformity

80% uniformity means that the weight of 80% of the individually weighed birds is between -10% and +10% of the average weight of all weighed birds.

15 weeks is a good age to determine uniformity. Weigh 1-3% of the birds spread over the poultry house and the different tiers. In a cage system, do this in each tier (feeding line). Even if there is a big difference in body weights, you should still start adapting the lighting scheme at the required average body weight. It will be clear that this is too early for the smaller birds and too late for the heavier ones. But there is no other choice. That is why uniformity is so important.

Factors influencing uniformity:

- Number of hens per m²
- Feeding structure (selective eating)
- Length of feeder and difference in height of feeder in several places
- Length of drinker (nipples) and availability of water
- Quality of beak trimming
- Stress factors (disease, vaccination)
- Age at which uniformity is measured (sexual development)
- Genetic background (breed of chicken)
- Weighing method: the more chicks you weigh, the more accurate the uniformity will be.

Automatic weighing in the house provides information about weight ánd uniformity.

Development of feathering

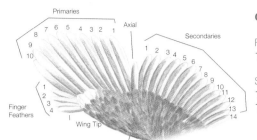

Source: Ron Jöerissen

Moult during rearing

Rearing hens moult four times: one complete moult and three partial moults. Feathers moult in a specific pattern. The first row of wing feathers is the most important for determining the moulting rate.

Order of moulting

First row:
1, 2, 3, 4, 5, 6, 7, 8, 9, 10

Second row:
11, 12, 13, 14, 10, 2, 3, 4, 5, 6, 7, 8, 9, 1, axial feather

Growth rate of wing feathers

A feather grows 75% of its final length in the first three weeks. The last 25% also takes three weeks. So a new feather is fully grown in six weeks.

Fully grown in six weeks

Last 25% growth takes 3 weeks

First 3 weeks: 75% of final length

Moulting stage of wing feathers

A smooth moulting process and a proper moult after 16 weeks is a healthy sign. At 15 or 16 weeks, count the number of wing feathers that have not yet moulted. Do not start light stimulation until all hens still have a maximum of two wing feathers to moult (moult score 2 or less).

Look at the primaries from outside to inside. They form a flowing shape in terms of length. This hen has moulted all its feathers. It is more than sufficiently developed to be able to start laying.

The feathers at far right are older and have not yet been moulted. This hen is still not sufficiently developed to start laying. The difference between feathers already moulted (rounded) and those not yet moulted (pointed) is clear to see.

Critical periods during development

Not all parts of the hen grow at the same pace; a different part grows stronger in each phase. In critical periods of rapid growth, chickens are particularly vulnerable. If something goes wrong then, their development will be impaired, resulting in the chicken coming into lay later or not achieving maximum production.

During periods of slower growth (roughly weeks 8-15), the chicken is able to cope better, and restricting its feed can in fact be a good idea. Also try and stimulate the volume of the feed intake during the last weeks of this period. You can do this by increasing the fibre content to reduce the nutrient concentration. Then they will eat more volume to meet their nutritional requirements. So they will eat more and their intestines will get used to a higher feed intake. Additional advantages of a higher fibre diet are: better digestion, drier litter, less ammonia, less feather pecking and better intestinal health. For the egg producer, the second critical period is particularly important (15-20 weeks) as this is the time when the flock is transferred: an extra reason to provide optimal care.

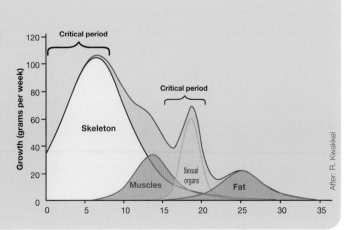

After: R. Kwakkel

Training good behaviour in aviaries

After layers are placed in an aviary, they find it easier to look for feed, water, laying nests and perches if they have learned to jump and look for things during rearing. You can improve hens' mobility by training them. This also reduces the risk of floor eggs. Also important for normal floor housing systems.

In a Nivo Varia house, as the hens get older more and more levels become available. Feed and water are provided further away from each other than in an aviary.

Aviary rearing: water training

The chicks are grown in the middle tier for the first few weeks. Everything they need is there: feed, water, fresh air and light. Not all chicks will leave this tier of their own accord: only the inquisitive ones will venture out. As soon as the chicks are physically able to jump, in other words when half or a third start jumping over the edge, let them go. Don't let the chicks stay on 'their' tier too long. After 7-8 weeks, force them to move by shutting off the water alternately on each tier, for example for a couple of hours in the afternoon. Thirst will make them go in search of water themselves and they will learn to move between the different tiers. Make sure no hens become dehydrated if they really can't find water.

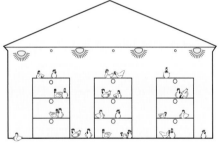

Daytime
- main lighting on
- rope lights off
- lure lights off

→ chickens in whole house

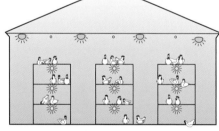

1 hour before dark period
- dim main lighting
- rope lights on
- lure lights off
→ chickens at the top stay there
→ chickens on the floor go to the bottom and middle tiers

1/2 hour before dark period
- main lights off
- dim rope lighting
- lure lights on

→ last chickens go into the tiers

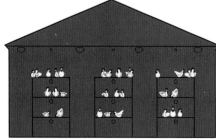

Dark
- main lights off
- rope lights off
- lure lights off

→ all chickens in system

Follow the light

To minimise the amount of manure in the litter and to maintain good litter quality, you want the chickens to sleep at the top in the system at night. Chickens are attracted to the light. Make use of this by operating the lights in the various compartments in the house separately. If you want them to be on perches or in the system in the evening, put the lights out in these places last.

You could provide additional dimming lights at the top of the house, for example. If you want the chickens to be in the scratching area during the day, make the light brightest there.

From rearing to laying: 16-25 weeks

The period from 16-20 weeks is puberty, a crucial and stressful time for the hen. The hen grows vigorously as her laying organs grow and fat is deposited. Minor errors can have major consequences at this time. So check that the feed intake and weight have increased enough and measure the body weight every two weeks. If a bird's weight stops increasing, her feed intake is too low.

You can stimulate the feed intake by running the chain more frequently, even for a very short time (block feeding). If you are using a feed trolley, you can turn the feed over manually or with a scoop. This will get them to eat more and their intestines will get used to a higher feed intake. If feed intake (g) remains low, supply more concentrated feed to give the birds more nutrients. An extra 30 minutes of light twice over two weeks can help. For flocks that already have a long day length at the end of rearing, you should increase the day length by an extra 1-2 hours during this period.

Temperature

Try to keep the poultry house temperature low at the start of the laying period (18-20°C). This encourages the birds to start eating (they eat 1.5 to 2 grams more for every degree lower temperature) and has a positive effect on the egg weight. To save feed, raise the temperature slowly to about 25°C as soon as the birds are eating enough.

While a pullet still feels bony at 10 weeks (photo left) and does not have much subcutaneous fat on her belly, by 20 weeks the hen will have developed a laying stomach with a lot of abdominal and subcutaneous fat (photo right).

Midnight snack

If the feed intake is not increasing rapidly enough, you can give your birds in a caged system a 'midnight snack' at about 23 weeks (but not if they are in a floor housing system as it results in floor eggs). This gives the birds an extra hour of light for no more than a couple of weeks in the middle of the dark period, which will encourage them to eat more. Stop giving the midnight snack as soon as their feed intake is adequate.

Insufficient calcium: cage layer fatigue

Give the birds coarse calcium in good time or they will take it out of their bones. If you provide calcium too late, around the peak production (25-26 weeks) you will often see a lot of birds whose legs give way or birds sitting on their hocks. This is also known as cage fatigue and results in a higher mortality rate. Make sure that the birds get extra calcium a week before they start egg laying. But don't increase the calcium content too much at the start of production, as it interferes with their energy and feed intake.

Hens that look as if they are in a daze, often with their eyes closed, are not drinking enough or not at all. Teach them to use the nipple or the water trough. Putting drippers on the nipple and moving these birds to the top tier with more light can help. Lower the nipple pressure to let them leak on purpose for a while.

Light

Make sure the light intensity for each bird is always higher when they are transferred. Birds located around the ventilator opening in the cage rearing house will generally already have had a higher light intensity there than other birds. The same often applies to the birds in the top tiers in the rearing battery.

When you transfer these birds make sure they go to the parts of the laying house with the highest light intensity. So put the birds that had been sitting right at the top by the lights, in the top cages in the laying house.

Give the birds constant light for two days after moving them so that they can find food and, most importantly, water. The intensity should be 10 lux or higher to stimulate their feed intake, particularly at the start of production.

Floor housing systems

When you move the birds to a floor housing system, make sure the laying nests are ready one week before laying starts. To prevent floor eggs, extend the day length during the morning hours and not in the evening. Collect floor eggs within an hour of laying to prevent the birds getting used to laying in the litter on the ground. This will always be in the early morning.

Tranquility and routine

Tranquility and routine are keywords in poultry rearing, just as with human children. A good lighting and feeding plan helps achieving this. The birds are encouraged to eat, but they also get sufficient rest. This helps reducing mortality. Each time the light goes on the chicks are activated to look for feed and water. This helps them synchronise their behaviour and get into a good rhythm.

You can control pecking by reducing the light intensity. With the minimum light intensity of 6 lux you will find it difficult to read a book in the house. But that is enough light for hens in the rearing stage. These days you can even get apps that can measure light on your mobile phone.

Light during rearing

Both the lighting programme and the composition of the feed and feed intake affect the growth and development of the chickens. With the pullets, start with the maximum light intensity during the first few days so that they can find feed and water easily. Then you can dim the light to keep the birds relaxed, which prevents cannibalism and impacts positively on the feed intake.

There are three phases:
1. In the first week full lighting is possible, so the pullets can easily find feed and water
2. During rearing the day length should never increase, but decreasing the light period is no problem
3. Increasing the light period will stimulate maturity and lay.

This is in fact the same as simulating spring (above 20° latitude), which is the natural season for laying eggs.

Lighting in dark poultry houses (black-out houses)

In this situation you can fully control the lighting. With the lights out, the light penetration through all kind of openings should be less than 0.5 lux. Maintain 10 hours of light (or 12 in summer time) during rearing. During laying this should be increased to 16 hours (at 50% of lay).

Golden rule
Never extend the length of the day and the light intensity during rearing until you want to start stimulating lay.

Dark poultry houses

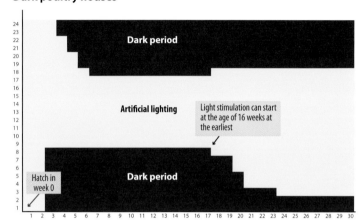

Natural daylight poultry house

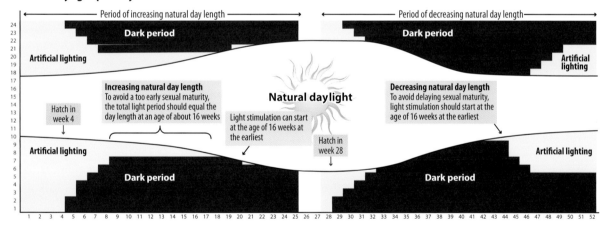

Lighting in natural daylight rearing

If you work with open sided houses, it is important not to increase the day length at all during the growing period (development and maturing phase). Establish what the natural day length will be when the birds are at least 16 weeks old. Then you can adjust the lighting programme to the daylight in two ways.

1. Provide additional lighting to make sure the day length from 8 weeks is the same as the natural day length at 16 weeks and keep it that way until it is time to stimulate production/sexual maturity. Often done when natural daylength is increasing.

2. Using additional lighting, provide a long enough day during the growing phase up to 10 weeks to enable you to shorten the day length in the last rearing phase so as to reach the natural day length at 18 weeks. This 'step-down system' will delay the start of production. Often done when natural day-length is decreasing.

Autopsy of a few pullets will tell you if there is already any evidence of chronic gastroenteritis. Medicines that are not permitted during the production period can still be used at the end of the rearing period.

Check that beaks have been properly trimmed, i.e. that the same amount has been removed from all birds. If it has not, as in the photograph, you may have problems with feed intake and wastage. The hen will also be less able to look after her feathers.

LOOK-THINK-ACT

Down feathers on the floor?

When you visit the grower, check whether there are down feathers in the litter. If there aren't, they have been eaten: a signal that the chicks are lacking in something, e.g. texture in the feed. Ask the grower to add alfalfa hay or another high-fibre product to the feed.

Moving from rearing to laying house

Move the hens from the rearing house to the laying house at the right time, at least two weeks before they are due to start laying. The transition from the rearing house to the laying house is a very stressful period for the hen, particularly if she is being moved from floor rearing to cage housing. Keep an eye on the hens in every cage at least twice a day for the first three days after moving them.

Make sure you know what has been going on during the rearing period. Go and take a look yourself from time to time. You should have information on vaccinations, feeding schemes, weight graphs, lighting schemes and age available. But don't forget to check other aspects like gizzard grit, a higher fibre content in the feed, litter, scattered grain, roughage and perches in case of floor housing.

Evaluation of the rearing period:

- Were there any health problems? What was the mortality rate (rearing list)? Were any birds removed, and why? How were they dealt with? Have the birds been given preventative treatment against worms? Are there red mites in the house?
- Were any particular actions needed, e.g. later introduction of hens? If so, why?
- What are the feeding and lighting times?
- Check the vaccination card to see how the vaccinations were done and ask about any irregularities.
- Check the results of blood tests.
- Physical development. Look at the size of the comb, the colour on the head and in particular: the moulting stage. Are the comb size and colour uniform? Pick chickens up in different places in the house.
- Weight development and uniformity. Are there a lot of small chicks? If so, why?
- Feather pecking. Are there signs of this?
- Light intensity. What light intensity were the birds reared in? Golden rule: the light intensity should be at least the same as or lower than in the laying house.
- Climate. Is there condensation on the walls? How are the chickens spread out in the house? Can you smell ammonia?

Vaccinations

Vaccinations prevent a number of diseases. Vaccination is a process in which the animals are infected with a much weakened or dead pathogen in a controlled way to build up a proper defence against infections.

Keep the number of vaccinations to a minimum, particularly during the first five weeks. Each vaccination produces a reaction, during which time the birds will be slightly sick for a few days. This stress may hinder the birds' development.

Only vaccinate the birds if they are healthy. This also means waiting a while between live vaccinations so that the birds are fully recovered from their previous vaccination reaction. Most vaccination reactions pass within 14 days. Check that the chicks are fit and healthy before you vaccinate them.

Less is more

Don't forget that a very intensive vaccination scheme will not necessarily offer more guarantees. Hindering growth before the fifth week of a chick's life can have serious consequences later on. Also bear in mind that a repeat vaccination can severely reduce the immunity that a chick has already built up to a particular virus for a short time, and the vaccination will not take well because of the chick's existing immunity. Repeating the same vaccination several times in quick succession is therefore risky in itself and is no guarantee for building up good immunity.

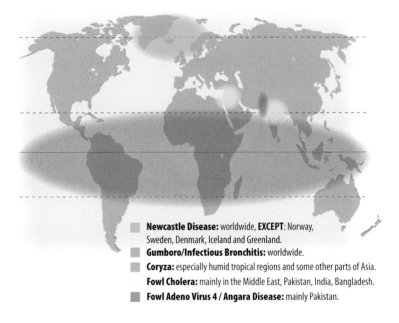

■ **Newcastle Disease:** worldwide, **EXCEPT**: Norway, Sweden, Denmark, Iceland and Greenland.
■ **Gumboro/Infectious Bronchitis:** worldwide.
■ **Coryza:** especially humid tropical regions and some other parts of Asia.
■ **Fowl Cholera:** mainly in the Middle East, Pakistan, India, Bangladesh.
■ **Fowl Adeno Virus 4 / Angara Disease:** mainly Pakistan.

Vaccination schemes differ between regions. Don't simply copy a vaccination scheme from another region indiscriminately. There is no point in vaccinating birds against infections that do not occur in your region; this can even impact on your birds' development.

Follow the immunity

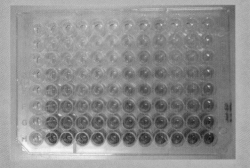

To get a good impression of your birds' immunity against a certain disease, you can have the titre determined. This will tell you how much resistance they have built up and how effective the re-vaccination will be.

A high titre is not always desirable. During rearing it is sometimes better to aim for a lower but adequate titre. This enables you to get a stronger vaccination effect just before laying and long-term protection during the laying period. Vaccination during the laying period can cause production problems.

If the titre of the immunity is too high during rearing, re-vaccination will have little effect. On the other hand, if the flock's immunity is too weak when they are re-vaccinated, a strong vaccination reaction during the laying period can also cause a drop in production.

No contact between laying hens and rearing hens

Older flocks in lay regularly catch infections but are not troubled by them as they have built up sufficient immunity. However, they can transmit the infection to rearing hens. Even a mild infection can impact the immune system, causing a carefully drawn up vaccination scheme to be disrupted. So keep young birds in rearing strictly separate from older flocks.

The right way to vaccinate

Good vaccination is a science in itself. Make sure that the people who are doing this are properly trained. To achieve a good result you also need to follow a few rules:

1. Use the right vaccine

For some diseases there is a choice of vaccines. To choose the right one you can have a lab test performed to find out which strain of the disease is present on the farm or in the area.

2. Store the vaccine properly

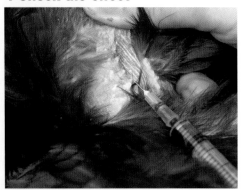

Vaccines must be kept in a cool, dark place. They lose their effectiveness very quickly if they become warm.

4 Check the effect

Take blood samples 4 to 6 weeks after the vaccination to test for antibodies in the lab.

3. Administer the vaccine in the proper way

Drinking water vaccinations
With drinking water vaccinations, all birds must ingest sufficient live vaccine particles.

Spray vaccinations
There are various devices available for administering spray vaccinations in the right way. Which device you choose depends on the number of birds to be vaccinated and what type of housing they are kept in.

Injections
Make sure the vaccine goes into the bird's muscle and not between the feathers, into the abdominal cavity or into the liver. Subcutaneous injection is recommended for some vaccines (particularly bacterial vaccines).

Reaction to vaccination: a good signal?

A reaction to a vaccination is a signal from the immune system that the vaccine has taken effect. After the first drinking water or spray vaccinations against Newcastle disease or IB, the chickens will be a little sick for a few days. You can tell the quality of the vaccination from whether all the chickens show the same reaction to it (good sign) or whether it 'rolls' through the flock (vaccination initially only affects some birds). Stress and a poor house climate (dust and ammonia) can worsen a reaction to a vaccination. If the birds are feverish, make sure the house is properly ventilated. You might need to give them extra vitamins and minerals in the water.

Points for attention when giving drinking water vaccinations

- Don't allow the chicks to drink 2-3 hours prior to vaccination, so that they are thirsty and soon empty the waterline containing the vaccine. If vaccine solution is allowed to stand for longer than two hours it loses its effectiveness.
- Prepare enough vaccine solution.
- Use a stabiliser that not only protects the vaccination virus but also has a distinctive colour: pepton, skimmed milk or special dye. Then you can tell when all the solution has been drunk.
- Check that the vaccine reaches the very back of the house by drawing it there.

Aviary rearing, vaccination misery?

Are the chicks still locked in the system, has the chick paper gone, and do they still need to be vaccinated against coccidiosis? Before these vaccinations take effect, it is important to recirculate via the manure or litter. Use thicker paper. Don't allow the chicks to eat before vaccinating them. If they are full and heavier, it is more difficult to get them into the system and they will become distressed more quickly if they sit on each other.

Signals of poorly performed vaccinations

This hen received a breast muscle injection. The injection needle was inserted too low, touching the liver. This animal is in shock.

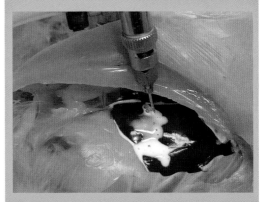

Here you can see what happens when you inject too low in the chest.

Minimise the stress at individual vaccinations. Prevent suffocation by providing sufficient ventilation and prevent the birds from crawling over each other. Treat the birds with respect and in an animal-friendly way. So don't handle them like the man on this picture...

Good vaccination - a science in itself

A vaccination plan is a guideline for the administration of vaccinations. But in order to achieve the desired effect and avoid adverse side-effects, the vaccination needs to be administered properly.

Right

Wrong

Make sure the equipment (filters, o-rings) is clean on the inside and outside.

Right

Wrong

For hygiene reasons, but also for your own safety, always wear gloves. Open the vaccine ampoule under water.

Right

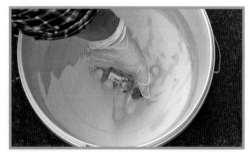

Wrong

Use a filter to prevent any sediment and other impurities from getting into the spray and blocking the nozzle.

Right

Wrong

If possible, use a vaccine containing a dye so you can see whether all the birds have taken it.

Keep the use of antibiotics to a minimum

Vaccination is prevention, but sometimes that is not enough. Antibiotics are an excellent way of killing bacteria, and sometimes they are unavoidable, but you should use them as little as possible. First try to improve housing and management to decrease stress and infection pressure so that you do not need to use antibiotics.

- The active ingredient in antibiotics not only kills the bad bacteria but also the good ones. This causes the animals a great deal of stress because they need the good bacteria for their intestinal and immune development.

- Not all bad bacteria are killed by antibiotic treatment; the strongest survive. These will multiply, producing ever stronger resistant strains against which the antibiotics will be useless.

- If animals are not 'infected' with bad bacteria, treating them with antibiotics will only have an adverse effect in that their good bacteria will be killed. Giving antibiotics preventively, for example in day-old chicks' drinking water, is detrimental. The chick's gut flora will be destroyed. In addition, some antibiotics turn drinking water bitter, making it unpleasant for day-old chicks to drink and causing them to drink too little.

- Antibiotics can, intentionally or accidentally, mask mistakes. Dirty drinking water can cause diarrhoea which necessitates the use of antibiotics. Keeping the drinking water clean avoids the need to use them, saves you a lot of money and prevents additional growth stress in the birds. The same applies to draught (causing colds and runny noses), unhygienic handling, and so on.

From small to adult in 18 weeks

Day-old chicks grow into sexually mature layers in just over four months. The development takes place on many different levels.

The last days in the hatchery

Before they hatch, the chicks 'talk' to each other to encourage each other to peck their way out of the shell. After breaking the shells the chicks are just about exhausted and rest while they dry off. All their reflexes are present straight away. Day-old chicks flap their wings as they drop off one moving belt onto the next in the hatchery. If they are lying on their backs, they turn over straight away. In the hatchery the chicks undergo several treatments such as sexing, occasionally beak trimming, spray vaccination against infectious bronchitis and/or Newcastle Disease and an injected vaccination against Marek's disease. All in all a good deal of stress.

Weeks 1-2

In the first few days they develop species-specific behaviour:
- they peck at everything to find out what is edible and to find something to drink;
- they eat their starter feed off chick paper or feeding plates;
- they learn to scratch in the ground, take dust baths and play;
- in floor housing systems: they learn to roost during the day;
- the gastrointestinal system develops during the first weeks.

Depending on the house, the poultry farmer will do the following
- aviary: chicks locked in the middle tier;
- house with slats; chicks shut up on slats and/ or in scratching area;
- beak trimming (if needed);
- vaccinations.

Weeks 5-6

Maximum physical development takes place in these weeks. All down has to be replaced by juvenile feathers (first moult). The skeleton is 50% complete.

Weeks 3-4

Chicks learn to feed from feed troughs or pans. Switch from starter feed to grower feed (phase 1).

Tasks for the poultry farmer:
- if available, provide litter for scratching behaviour;
- vaccinations.

Weeks 7-8

During these weeks, hens are eager to be fed: let them eat all the feed in the chain feeder once or twice a day to prevent selective intake. Sometimes there will also be a partial moult which makes the chicks more susceptible to problems. The chicks fight to establish a (temporary) pecking order.

Egged on from all sides

The changes a hen undergoes during her development are partly genetically preordained: which organ grows and when, which behaviour is learned and when, and when the feathers are shed. Many external influences also affect her development. She is prepared for the production phase via the lighting programme and the composition of the feed. Other influences include the vaccination plan and beak trimming. In the long term these treatments increase the hen's chances of survival, but in the short term she has to deal with stress, pain and reactions to vaccinations.

So every development stage has its own points for concern for the poultry farmer.

Weeks 12-14

The second part of the partial moult starts in weeks 12-13. It continues until week 16. Interruptions in the moulting process can be seen straight away in the replacement of the wing feathers. Stress factors and vaccination plans affect the start and end of this moult. Oviducts and ovaries start to develop. Immunity continues to build up.

Weeks 17-18

The young hens are transferred to the laying farm.

Photo's: Interbroed layers

Weeks 9-11

The birds are now almost fully grown in terms of size, but not in terms of muscle development and fat deposition. A partial moult takes place during this period. Body feathers are replaced but tail feathers remain (second moult). Provided the flock is sufficiently uniform and is above the normal weight (i.e. not on the basis of age), you can start slowing down the increase of feed supply. To prevent unnecessary high nutrient intake, switch to a phase II grower feed with a lower nutrient concentration. This approach will result in efficient birds in the laying period.

Weeks 15-16

A young hen is at its most vulnerable in weeks 16-20. The skeleton is 95% complete. The laying organs begin to develop now. This means a rapid increase in body weight: this is caused by fluid retention and fat deposition rather than higher feed intake. The sound changes: the hens begin to cluck. The hens get colour on their heads and the combs get bigger. The final vaccinations are administered. Do a final check of blood samples for antibody titres, weight and uniformity. Check the wing feathers to make sure the moult has proceeded properly, because this is an important signal as to whether the hen was raised properly.

Feather pecking during rearing

Feather pecking among rearing hens is greatly underestimated, and the impact of this on feather pecking in the laying period is greater than is thought. There are various signs of feather pecking at an early age:

- Feathers lying on the ground disappear (they are eaten).
- You can hear cries of pain every now and again (when a feather is pulled out).
- You can see injuries on the hens. Warning: Feather pecking among chicks is much more subtle than among adult chickens. You will rarely see bald patches on chicks.

Risks and prevention

- Boring environment. Chicks naturally peck the ground. If there is nothing of interest lying around (feed, litter), they will start pecking other things. Prevention: feed on feed plates or paper as long as possible and provide litter.
- Let the chicks out of the system or off the slats as early as possible.

- During moulting periods, chickens have untidy plumage. This encourages them to pull out feathers.
- Lack of water and feed, feeding structure not right, selective eating, need for fibre causing nutrient deficiencies, empty intestines and hunger feeling. So they pull out each other's feathers. Prevention: give fibre in the feed, alfalfa hay or extra litter.
- Redirected social behaviour. In large groups they always encounter strangers whom they peck to get to know. Prevention: lower occupancy or smaller flocks.
- Poor house climate (carbon dioxide, ammonia, hot and dusty) and high light intensity.
- Beginning of laying, when the vent membrane invites pecking.
- Beware of a high density. The chicks should have enough space.
- In the first week, offer perches or other elevations so that they can flee from each other.
- Lice.

Recognising pecking

If at the age of 16 weeks 20% of the birds are displaying symptoms of feather pecking....

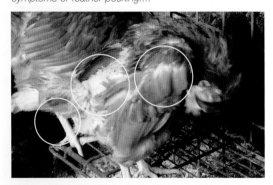

... all the chickens in the flock will have bald backs by the time they are 35 weeks old.

If you put the chicks on wood shavings from day one, as in this photo, there is less chance of feather pecking.

Beak trimming

Beaks are trimmed to prevent pecking. Pecking is more likely in large cages and with high density. Beak trimming can prevent this, but it is a major intervention on the bird. It is least stressful when done on day-old chicks, particularly if it is done using an infrared device. But there is a greater risk of increased mortality during the first few days if not done correctly. Beak trimming involves removing part of the beak. The wound is cauterised immediately with the heated blade. At an early age the intervention is less painful for the chick and the risk of chronic pain is small, as long as no more than one third of the beak tip is removed. Beak trimming is usually done between 7 and 10 days.

Advantage of beak trimming

+ Chickens are less able to feather peck. The damage caused by feather pecking is less.

Disadvantages of beak trimming

− Trimming is painful and stressful, and some birds are left with chronic pain.
− Bad for the public image of the sector, and costs money.
− Can cause permanent damage to the bird if done badly.

Example of incorrect trimming: only the top beak has been shortened. Badly de-beaked chickens waste more feed and find it harder to eat.

Spectacles?

Instead of beak trimming, you can also use plastic 'spectacles'. These spectacles are attached to the nose with a pin through the nostrils at around six weeks. They stop the birds from looking straight ahead. This method is very effective and costs no more than beak trimming. Spectacles cannot be used in cage housing because the spectacles can get hooked in the cage walls, trapping the birds. But is this an acceptable option in terms of animal welfare....?

Precautionary measures when beak trimming

1. A few days before beak trimming, give the birds extra vitamin K to promote blood clotting.
2. Do not beak trim birds if the flock is not in good health or if it is suffering from vaccine reactions.
3. If beak trimming at a later age, do this at the coolest time of day and never during very hot weather.
4. Make sure the birds waiting to have their beaks trimmed are hungry and give them something to eat immediately afterwards so that they look for food straight away. Contact with food can limit bleeding.

Performing beak trimming

1. Check the equipment and make sure that the trimming blade is at the right temperature for cauterising (cherry red colour) but is not too high (risk of blisters on the beak).
2. The operator should be installed and seated comfortably so that each beak is cut in the same manner.
3. Handle the chicks gently. Make sure the tongue of the bird does not get burned. After trimming, drop them gently onto a box with wood shavings to prevent injury.
4. Do not rush the process: too high a rate (number of birds/minute) could lead to a higher chance of errors and poor uniformity.
5. Clean the blades with sandpaper after use on 5,000 chicks, and renew them after 20,000 to 30,000 chicks.

Aftercare

1. Increase the water level in the drinkers and if using nipples, lower the pressure to make it easy for the birds to drink.
2. Make sure the feed level in the feed hoppers is kept high for a few days after beak trimming.

Methods of beak trimming

Beak trimming should always be carried out by properly trained personnel. When improperly done it results in:

- birds having difficulties eating and drinking
- unevenness in the flock, poor performance and more mortality.

Apart from regular beak trimming (heat), the infrared (IR) technique is becoming increasingly widespread. This still causes acute pain, but there is no open wound, so there is less bleeding and infection. The pain during drinking and eating in the days after treatment is reduced.

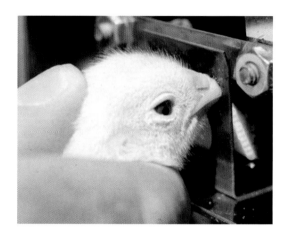

Beak trimming at the hatchery

You don't have to catch the day-old chicks since they are going through a series of treatments anyway. So this is a very labour efficient method.

Beak trimming at 7-10 days

Choose the correct diameter in order to cut the beak at least 2 mm from the nostrils. Hold the chick in one hand: thumb behind the head, index finger under the throat to apply a slight pressure back and downward to hold the tongue back. Tilt the beak at a 15° angle and cauterise for 2-2.5 seconds. Check the temperature of the blade (600-650°C) for each operator and machine every hour.

Results of beak trimming at the hatchery

OK

Not straight

Too hot and too short

Too hot

Not straight and too hot

Too little

Beak trimming at 7-10 weeks

Because this is painful for the hens, beak trimming at this age is not permitted in several countries. Avoid this if possible. If necessary to do

- Insert a finger between the 2 mandibles (preferred technique).
- Cut the beak perpendicularly at a right angle to its long axis, so that about half the length of the beak between the tip and the nostrils is left after cauterisation
- Cauterise each mandible with care, particularly at the sides of the beak, so as to round off the sides of the beak and avoid lateral re-growth
- Check the temperature of the blade regularly (650° - 750° C)

Objective of trimming and the result of good trimming in adult hen.

Beaks should preferably be cut shorter in an open, well-lit house (left) than in a dark house with no daylight (right).

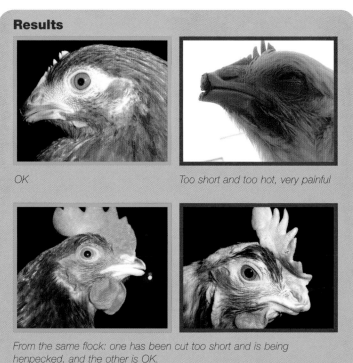

Results

OK

Too short and too hot, very painful

From the same flock: one has been cut too short and is being henpecked, and the other is OK.

Big difference in quality of beak trimming: the bird at far right has been done well, the others have not.

Beak trimmed at 6 weeks. De-beaking retards growth by 10 days and there is a high risk of chronic pain. This is now no longer permitted in some countries.

CHAPTER 4

Laying hens

Laying percentage and egg weight depend on the breed of hen and your management. The ideal weights for eggs for the table differ from country to country, but are often sizes M (53-63 g) and L (63-73 g). In some countries XL is also fine provided the shells are of good quality.

Eggs are the main source of income on a layer farm. But all sorts of things are involved in achieving a high laying percentage and good quality eggs.

Schedules aren't written in stone

Don't apply schedules blindly as circumstances can differ quite considerably. Look in particular at the development of the chickens. For example, provide the chickens extra light if they have moulted completely and when they have reached the right body weight. If they have been reared well, this varies between weeks 16 and 18, but as said this can vary. You want all hens to come into production at the same time. So make sure that as many as possible have the same body weight (good uniformity) and are in good condition when you move them to the laying house. Keep in regular contact with the grower. If there are a lot of hens at the same stage of development, you can manage them well with feeding and lighting programmes. So fewer hens will come into production too early or too late.

Flocks that come into lay too early are often poor flocks that never reach peak production and are worn out before their time. They eat too little feed, so their body weight is too low, the weight of the eggs is too low, the quality of the eggshells is poor, and they have poor laying persistence, higher mortality and a greater risk of bad feathering. The latter can sometimes even lead to almost featherless chickens early in the laying period.

Tip

An extra week of production at the end gives you more benefits than a week shorter rearing period. Remark: hens kept in floor housing systems should have sufficient time (1 week) to get used to their new environment before laying starts. This is less of an issue in cage systems.

Tips for looking after the new arrivals

- Know how many hens you are getting and know their age.
- Make sure that there is food and water in the laying house before unloading the hens.
- Put the new hens in the laying house in the morning, ideally near the feeder and drinker. You might like to leave the lights on longer on the first day. Two days of constant light is best in cage systems.
- Make the lighting the same as in the rearing house after the first day. To prevent delays in production, the light strength and day length should be no less than at the end of the birds' time in the rearing house. A slight increase in light intensity (spring) is recommended. In a cage system, keep the light intensity high for the first few days.
- Do not switch from coarse to fine feed.
- Encourage activity and movement by walking through the house at irregular times. In an aviary system, for the first few days after the chickens arrive chase them off any levels without feed or water to prevent individuals from going hungry or thirsty.
- In the evenings, go into the house and pick up the chickens off the ground and put them into the system. This also prevents hens from laying floor eggs.
- Use step slats to make it easier for the hens to get in and out of the system.
- Make sure that the house is at the right temperature (ideally 18°C).
- Spread a thin layer of litter to begin with to prevent floor eggs
- Examine the weight and the uniformity of the hens carefully. What stage of moulting are they at?
- Release the hens in the system or let them get out of the crates themselves.
- Check the distribution of the animals in the poultry house.

Moving in

Moving to a new house is very disruptive for the birds, so it is important to do all you can to help them settle in quickly. Think of it like friends coming to visit: you would also offer them a nice cup of tea and a biscuit in a comfortable room. The same applies to how you welcome your new hens. After a long journey they are arriving at their new home. Make it quiet and comfortable.

For example, make sure they can find everything easily and that the temperature is right. The better you do this, the more likely you are to have a good production cycle and the less likely you are to have problems. In effect, you are simply continuing to rear the birds. You only become a laying hen farmer when the first egg is laid; until then you are a grower. Don't transport the birds with full crops, but do give them a constant supply of water.

18°C is a comfortable house temperature. Warm up the poultry house if necessary before the hens arrive. After all, the hens have not eaten for a while and will get cold more quickly, risking getting off to a poor start.

The drinking water system in the rearing house must be similar to the system in the laying house. If you have a nipple drinking system, make sure the drop is visible so the birds recognise it as a water source. The colour of the nipple could play a role as well, so you could also use some nipples in the 'rearing colour'.

Watch the weight development of the hens. During the first 10 production weeks, weigh hens once a week at the same time of day. Automatic weighing is even better because it does not cause the birds stress.

The ideal curves

Egg production is well under way at around 20 weeks and the peak laying percentage is achieved at around 28 to 30 weeks. The egg weight should grow quickly to 60 g at 30 weeks, after which it will increase gradually. Follow the laying percentage, feed and water intake and egg weight carefully, and intervene if production lags behind or starts to dip too soon.

A uniform flock of sufficient weight gives the best chance of eggs of the right weight at the start of the laying period and is also easier to manage. You can also stimulate the hens with light slightly later, so the first eggs will be laid later and will therefore be heavier.

If the egg weight or production stays low, check the health of the flock, the feed intake and the feed quality.

If disease is not a factor, adjust the feed and the lighting programme. Discuss this with your feed supplier. For example, more methionine and linoleic acid can influence the weight of the egg. Make sure the eggs do not get too heavy (once they are too heavy it is difficult to reduce the weight without losing laying rate). After peak production, luxury consumption may occur. You can then restrict the feed, but take care because of the risk of undernourishment and feather pecking.

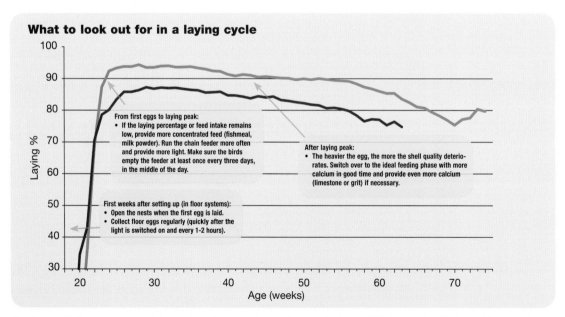

What to look out for in a laying cycle

Laying % vs *Age (weeks)*

From first eggs to laying peak:
- If the laying percentage or feed intake remains low, provide more concentrated feed (fishmeal, milk powder). Run the chain feeder more often and provide more light. Make sure the birds empty the feeder at least once every three days, in the middle of the day.

After laying peak:
- The heavier the egg, the more the shell quality deteriorates. Switch over to the ideal feeding phase with more calcium in good time and provide even more calcium (limestone or grit) if necessary.

First weeks after setting up (in floor systems):
- Open the nests when the first egg is laid.
- Collect floor eggs regularly (quickly after the light is switched on and every 1-2 hours).

The flock with the green line is doing well, as the curve has a clear peak above 90%. The flock with the red line is doing less well. A poor peak and a too low laying percentage for the whole period which could not be prevented. Up to week 22 everything seemed fine...

Gear management towards nesting peak

Hens are laying for ever longer periods: sometimes up to more than 100 days in succession. The first egg in a series is laid early in the morning. With present-day breeds the laying time hardly ever changes, while with earlier crosses the laying time would change and the hens sometimes missed a day. Most eggs are therefore laid in the morning. Gear your management towards this morning laying peak. A hen needs to be able to lay her eggs in peace. Make sure feed and water is available when the lights go on in the morning. Then let them rest and don't feed them for five to six hours, until most of the eggs have been laid. In the late evening run the chain feeder for two hours before the light goes out so that they can eat well before they go to sleep and still have enough feed in the morning. In the morning, don't run the feeder until after the laying peak. This prevents the birds being lured out of the nests, causing disruption and more floor eggs.

Fear costs eggs

Make sure the hens get used to people. Fear of people will affect egg production and egg quality. Genuine panic reactions can even lead to higher bird losses and a higher risk of feather pecking due to the stress. It also makes it more difficult to perform health checks and catch hens. Put a radio on in the house, ideally with a programme that alternates music and speech. This will get the hens used to human sounds so they will take fright less easily.

Tips for getting chickens used to people:

Pop in to the hens several times a day, not always at the same time. Wearing different colour overalls helps as well.

Regularly scatter small amounts of grain by hand. The hens will associate the 'care giver' with a positive experience.

The more intriguing and richer the surroundings, the less anxious the hens will be in general. In some countries: consider building a winter garden with perches and a roughage dispenser.

LOOK-THINK-ACT

What does this mean?

This is an anxious flock. All the birds are trying to get away and they are piling up on top of each other. This could result in trampling and suffocation. Frightened hens in cages can't run away but they could injure themselves seriously. If hens are easily frightened, try and get them more accustomed to people by walking through the poultry house more often.

Feed selection

Chickens prefer the coarse components of their feed. But the finer components also contain essential ingredients. Selective intake can be controlled by checking whether there are still enough coarse components in the feeder after 10 minutes. Different sized components must be evenly distributed to avoid selection. Tailor your feed management towards ensuring that all the chickens take in the same nutrients: the most dominant hens will make a bee-line for the coarse components. A fast-moving chain can prevent birds at the front of the chain picking out all the coarse components. The type of feeding system also has an impact.

When are the component sizes badly distributed?
1. Indication: What's left in the feeder after 10 minutes
2. Observe: Demixing in the silo
3. Sample feed at beginning and end of the chain for chemical analyses.
4. Talk to the feed supplier about this.

To stimulate the feed intake of a poorly reared flock in terms of body weight, add 1-2% fishmeal or adjust the feeding schedule for a week

Lighting and feeding programmes

The lighting programme plays a major role in the development of the animal and its later laying performance. The basics are simple: a decreasing day length or continuous short day inhibits the natural sexual maturity and a increasing day length stimulates this (also see page 59).

Assessing the structure of the feed

The structure of the feed can be measured with a sieve analysis or by visual observation. Sometimes the structure will look reasonable or good, but the chickens' feed intake will still be very selective. In the trough you can see the extent to which the chickens are feeding selectively.

Grower feed with a good structure. There is relatively little difference between coarse and fine components.

Coarse components are still visible even in the last bit of feed in the chain feeder. All animals are taking in the same amount.

All coarse components have disappeared from the feed. The hens have evidently been looking for coarse components. And especially the dominant ones will have eaten the maize and wheat.

Feed and light

With feed and light you influence development, production and behaviour. In the rearing period good management transforms a small one-day chick into a beautiful laying hen.

Phase 1: 0-6 weeks

During this period the birds need to grow strongly and therefore take in ample amounts of nutrients. For the first week the chicks need to eat mainly easily-digestible feed. Sufficient maize improves digestibility. From three weeks on, they have to learn that they won't get feed the whole day. So when the feed is finished, it is finished! But only restrict feed if body weights are on target. During this period the moulting from down to feathers should be completed. Furthermore, this period is an important basis for a good skeleton.

Phase 2: 7-15 weeks

This is the period in which the body weight can be influenced without any adverse effects. A higher body weight enables production to get off to an earlier start. A lower body weight means a later start. Whatever you do, just focus on uniform development. In this period there are two occasions on which animals will go through a partial moult. These are from week 7 to 9 and from week 12 to 16: always important phases in their development. Pay extra attention during these vulnerable periods.

Phase 3: 16-19 weeks

The hen has to get ready for laying. At this stage the birds have to be strongly encouraged to eat in order to grow well so that the laying organs will develop properly. At the same time there is another partial moult from week 19 to 21. This is a very critical period in which much can go wrong, resulting not in beautiful birds but poorly performing ones.

Phase 4: 20-30 weeks

During this period the birds reach their laying peak, but they are not yet fully grown. Ensure that the birds keep growing, but not by providing feed the whole day. They should stay eager for feed and should not waste any. You accomplish this by making sure that the feed level in the hopper is kept low and that the feeding system is emptied completely at least once a day. Just like in the rearing period.

Phase feeding for layers

Layers have a different need for energy, protein etc. in each phase. Phase feeding enables you to control the egg weight. You will also feed less protein (less N) throughout the production period, so you will save money. There is no such thing as a standard recommendation. The ratio of feed intake to grams of egg production is an important yardstick, however. If pullets arrive on the laying farm early (16 weeks), you can opt to continue feeding rearing feed for another week or start the pre-lay feed or starter feed straight away, depending when you expect production to start. These feeds have a higher nutritional value and are designed for growth and development of the laying organs: essential for laying percentage and egg mass. At 19-20 weeks you can switch to laying feed if the laying percentage is about 5.

Development visualised

Growth is not the same as development. Development continues even when the skeleton is fully-grown. And the hen has different needs for things like muscles and fat than for skeleton growth.

A slim hen at 10 weeks, with full plumage. This hen will not grow much more in terms of length. The skeleton is almost fully grown. When you touch the skin at the breast of this hen, you will feel almost no fat.

A big hen at 22 weeks. Starting at 10 weeks, this hen has developed quite a muscular chest, with a thin layer of fat under the skin. The chicken's round belly shows the development of the laying organs, with more fat.

Diluting the feed with roughage

The more time chickens spend feeding and foraging for food (scratching, scraping the ground and pecking), the happier they are. In a floor system you can encourage this behaviour with roughage (e.g. alfalfa hay) or loose grains in the scratching area. In a cage system you can dilute the laying feed with ground roughage as soon as the peak period is over. Chickens compensate for the diluted feed by eating more of it. Diluting by 10-15% has no adverse effects on production. Supplementary feeding or diluting usually produces healthier chickens, lower mortality and less risk of feather pecking.

Scattering grain activates their natural behaviour and encourages the hens to disperse well throughout the poultry house.

Deficiency signal: eating feathers

Are there down feathers in the litter? If the hens are eating the feathers, there is stress in the flock. This stress causes the hen's gastrointestinal system to malfunction. The bird wants to compensate for this by eating structure-rich components, including feathers.
This stress can be caused by:
● house climate
● lack of structure in the feed
● shortage of nutrients and crude fibre
● infections.
Often it is a combination of these factors. Always involve your feed supplier and change the feed composition if necessary. For example, provide more high fibre products. Don't forget to consult the vet.

Feeding when temperature is low

The optimum ambient temperature for laying hens is about 25°C. If the ambient temperature is very different from this, you will notice it in the birds' feed intake.
If you notice feed consumption increasing, this may be because the poultry house temperature has dropped to below 20°C. The hens start to eat more because they need more energy when the temperature drops this low. The birds take in about one gram more of feed for every degree Celsius. At some point it could become so cold that they can't take in enough feed. Then you will need to give them concentrated feed or they will start laying less.

Feeding when temperature is high

If you notice that the birds are eating too little, this can have a number of different causes. And one of these is that the temperature in the house is too far above 25°C. If the hens start eating less, they will not only take in less energy but also less protein and minerals. This affects the start of production, especially in older rearing hens.

A protein deficiency in laying hens will soon result in lower egg production and egg weight and higher mortality. The shell quality is also poorer because of the increased excretion of minerals, higher blood pH and lower calcium intake. They also need more vitamin C than they can produce themselves. Modified feeds can help, as can adding vitamin C to the drinking water. Particularly if the temperature rises above 32°C, the birds should be given special feed with a high energy value, preferably by adding fat and, depending on production, not too much protein. Excess protein intake produces a lot of heat during digestion.

Heat stress: increased water consumption

A hen normally drinks about 1.8 times her feed intake. This means about 200 g water (200 ml) per day. If the temperature rises above 30°C, the water intake will increase very sharply because the birds evaporate a lot of water through their breathing. Water intake can then rise to 300 ml or more per bird per day. This means that a bird with a body weight of 1500 g takes in about 20% of her body weight in water every day. Drinking water intake can help the hen reduce the stress caused by the high temperature.

Water consumption (litres/1000 birds)

	20°C	32°C
Pullets		
4 weeks	50	75
12 weeks	115	180
18 weeks	140	200
Laying hens		
50% prod.	150	250
90% prod.	180	300

These water consumption figures are indications (also depends on relative humidity, health, feed consumption, etc.), but it is clear that a 50% increase due to heat stress is not uncommon.

Cool water

Help the chicken in high temperatures by ensuring that the water is cool:
1. Make sure the water supply (water tower) is in the shade and is well insulated.
2. Make sure the supply pipe to the water tower and the water pipes to and inside the poultry houses are not in the sun, well insulated and preferably underground.
3. Allow the water at the end of the drinking lines to flow slowly.
4. Put ice blocks in the water tank if temperature is too high.

This water tank is heated up by the sun all day long because it is so high up (to achieve enough pressure). So the water is warm when it enters the house...

A roof over a water tank keeps it out of direct sunlight and stops the water heating up.

What to look out for in feeding systems

Make sure the feeder is long enough so that all the birds have a chance to feed at the same time. Taking account of the size of a laying hen, in a chain feeder system each bird needs a feeding length of 15 centimetres to allow them all to feed at the same time. If hens are reared in a different feeding system than the one used in the laying phase, make sure they are not too frightened by the new system.

Trough with chain feeder

A chain transports feed from a storage bin and distributes it along the feeder trough. The speed varies between 4 and 20 m per minute. With a faster chain there is less risk of selective feeding. A grill above the feeder troughs prevents hens from crawling into the troughs and contaminating the feed with droppings. Also check the height of the feed. If the layer is thicker, more will be spilt and the birds will find it easier to feed selectively.

Trough with spiral

The spiral distributes the feed over the trough and stops the hens throwing it to the side, so they spill less.

Feeding pans

Feeding pans are height adjustable. Make sure the feed is evenly distributed over all pans. Spillage can be a problem, but it depends to a large degree on how the system is installed. The pans must be hung at the right height, low over the litter (at height of the back of the hen). It is important to feed in blocks, meaning that all feed hoppers are filled in a short time.

Tip

Take a sample from every feed delivery. If something goes wrong later, you can always check whether it has anything to do with the feed. Keep the sample for at least four weeks.

Drinking

One cause of young chicks drinking too little is that the nipples are too high or too low. Change the height and see whether this affects their water consumption. Hens need to be able to drink easily. Otherwise they will drink too little, eat too little and they won't grow properly.

Water intake

Water intake depends on feed intake, feed composition, house temperature and age. As a rule of thumb, from 10 days the ratio between water and feed should be 1.7. Layers drink little and often, altogether about 200 ml a day. The volume of daily water intake is a good indicator of the health of the flock. Record the daily water and check feed intake. A sudden change in water intake is an important signal. If the water consumption increases, first check for leaks in the drinking system and check the water pressure, the temperature in the house and the salt content of the feed. If this is not the reason, check the birds' health (disease, vaccination reaction). Also check whether the change coincides with a feed supply or a change in the feeding phase.

If the hens are drinking too little, first check that the water system is working properly. The water pressure must not be too low, otherwise water will leak out. You also don't want the water pressure in the drinking line to be too high, as this can cause lower water intake because the hens have to push harder on the nipple. Hens that drink too little look drowsy. Check that the nipples in all cages/areas with drowsy birds are working properly. Check the water quality when the system is working properly and the nipple is at the right height.

Drinking water must be readily available. In traditional battery cages, 2 drinker nipples or cups must be available. The standard for hens in floor systems is 1 drinker nipple or cup per 10 birds or 1 cm per bird (round drinkers), or 2 cm with long drinking troughs. These days drinker nipples are of such good quality that drip trays are no longer necessary. Drip trays are susceptible to fouling.

Pros and cons of drinking water systems

Round drinkers	Drinking nipples	Drinking cups
+ Water is readily available	+ Closed system, water always fresh	+ Water is readily available
+ Water level and suspension height easy to regulate	+ Very little spillage	+ Easy to check for blockages
− Open system, not always fresh, more chance of contamination	+ Lots of room to walk around	− High investment costs
− Spills, wet litter	− High investment costs	− More chance of contamination
	− Water dispensing harder to control	− Less room to walk around

LOOK-THINK-ACT

Dry litter = good litter?

What do you think of the litter in this picture? You can see that there are particles fluttering down to the ground. This is a good sign because it means that the litter is dry. However, litter always gets slightly wet from leaking nipples and spills. If the litter is too dry, this might be a sign that the chicks are not drinking enough. Check the water intake and if necessary the water output from the nipples all over the house.

Water output from the nipples

If there is too little water coming out of the nipples, the chicks will drink too little. Check the water pressure and the water output from the nipples regularly. You can assess the flow speed by holding up a container to a nipple for one minute. Measure the amount of water in the container. Do this at several different drinking lines. A handy rule of thumb for the flow rate is the age of the chicks in days plus 20 ml/minute. For example: 35 days + 20 = 55 ml/minute. Too much water will cause spillage and bad litter quality, which will in turn lead to poorer quality chicks and footpad lesions. Have the drinking water checked by a laboratory. Check the drinking lines visually for soiling. Do this on the inside and outside of the system.

What is the right drinking position?

The right drinking position is upright with the head up so that the water runs into the throat. What is upright? You can control this by raising or lowering the drinking nipples.

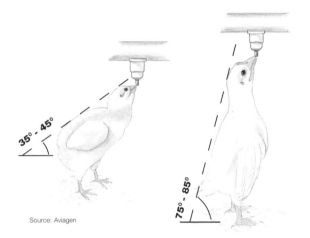

Source: Aviagen

For one week old chicks, the angle between the beak and the nipple should be 35-45°.

For chicks older than one week, the angle between the beak and the nipple should be 80-85°.

Check the water:

1. generally on the meter (automatically)
2. per nipple line: the level at the air vent (daily)
3. per nipple (at least once every two months)

Water quality

Water can contain dangerous, undesirable substances that will end up inside the chicken. The quality of mains water is generally good. But this may not be the case with water from your own source. In order to keep your water quality up to standard, technical aids such as a deionisation system or a reverse osmosis system (which filters substances out of the water) can be used.

Signal	Cause	Dangerous (mg/l)
Lower oxygen uptake in the blood, result: comb, lobes and head turn blue, animal becomes lethargic. Reduced fertility	Nitrite	> 1,0
Respiratory problems	Nitrate (can be converted into nitrite)	> 200
Diarrhoea Cerebral symptoms: wryneck and lameness	Sodium	> 200 [1]
Reduced feed intake	Chloride	> 300 [1]
Blocked nerve conduction; smell of rotten eggs	Sulphide, conversion from sulphate under the influence of certain bacteria	> 250
Intestinal dysfunction	Iron	> 5,0
Intestinal dysfunction, various bacterial problems	E. coli	> 100 (kve/ml)
Weaker resistance	Mycotoxins (produced by moulds)	No threshold

[1]: for laying hens > 600 mg/l chloride and > 400 mg/l sodium

Check the water quality in the house

The drinking water should taste pleasant and must not contain any hazardous substances or impurities. Additives such as antibiotics can end up in the eggs and pose a food safety risk. Water also serves as a solvent for medicines and vaccines. When vaccinating via the drinking water, make sure the water is clean and cool and that the pipe is working properly. So rinse the pipe through well beforehand. Rinse the pipe out thoroughly again afterwards to prevent residues.

Adding antibiotics or medicines to the water can give a bitter taste, so the birds will drink less.

Also clean water pipes well and avoid sagging to prevent fungal growth. If you suspect the drinking water might be contaminated, have it tested. Always check the quality and temperature of the water at the beginning and end of the pipes. You often find that good quality water is not so good at the end of the pipe.

Testing the waters

You can get a quick initial impression of the water quality by pouring some water into a glass jar and examining the colour, clarity, sediment and smell. Also check whether there is a membrane floating on the water. Take the water from the drinking points at the front and back of the house. It's easy to assess; every parameter is rated good, moderate or poor.

Colour: good (absolutely colourless); moderate (slight discolouration); poor (distinct colour; yellow, brown etc.).
Clarity: good (completely clear); moderate (cloudy but still transparent); poor (opaque).
Sediment: good (water free from particles); moderate (a few particles); poor (bottom is entirely covered with mud or iron particles).
Smell: good (absolutely odourless); moderate (slight smell); poor (strong smell of rotten eggs).

Assessment:
- All parts good: 15% chance that the water is unsuitable.
- All parts poor: the water is always unsuitable.
- Some parts moderate: 35-75% chance that the water is unsuitable.

Test the water at least twice a year. Sample it at the end of the nipple line or the end of the system. Also ask yourself: would I drink this water myself? If not, why would your chickens?

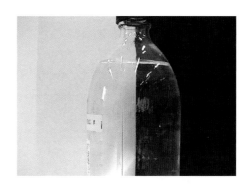

colour: good, clarity: good

colour: poor, clarity: good

colour: poor, clarity: poor

colour and clarity good; sediment poor

Disadvantages and advantages of forced moulting

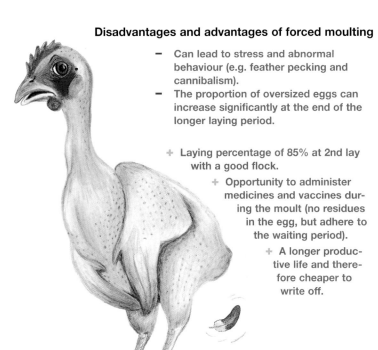

- − Can lead to stress and abnormal behaviour (e.g. feather pecking and cannibalism).
- − The proportion of oversized eggs can increase significantly at the end of the longer laying period.

- + Laying percentage of 85% at 2nd lay with a good flock.
- + Opportunity to administer medicines and vaccines during the moult (no residues in the egg, but adhere to the waiting period).
- + A longer productive life and therefore cheaper to write off.

A second laying period?

Hens are laying for longer these days. There is less reason to keep them for a second laying period. In emergencies or unforeseen circumstances, it is sometimes decided to force a moult. The moult lasts for four to six weeks and the second laying period lasts for six to eight months. During the moult the birds' weight will reduce by 30%, mainly as a result of using fat. Forcing a moult is a skilled operation. Consult your advisor and abide by the statutory requirements. Give extra grit while eggs are still being produced, to prevent the birds from becoming deficient in calcium and suffering skeletal defects.

Pecking, feather pecking and cannibalism?

Pecking starts at the tail joint and continues until blood is drawn. It almost always happens to young birds that have just acquired their plumage. Remove the bird and spray it with something with an unpleasant taste and smell, such as hartshorn oil, so the other birds no longer want to peck at it. If you don't intervene fast and effectively, it will develop into a major problem.

Feather pecking among rearing hens is underestimated. On adult hens you could see bald patches, but in pullets you often only notice a few covert feathers missing at the bottom of the back. You can recognise this by the protruding down feathers or bushy tail feathers. It is more noticeable on brown hens than on white ones as the white underfeathers are under the brown coverts. Genuine bald patches are very rare in the rearing phase. If a few underfeathers are visible on 20% of the chickens at 16 weeks, most of the flock will have significant bald patches by 30 weeks.

Types of pecking

Scientists distinguish between two types of pecking: 'aggressive pecking' and 'feather pecking/cannibalism'. The signals from the chicken differ for each type. To be able to respond appropriately, you need to be able to recognise the signals. Feather pecking is often incorrectly described as aggressive behaviour. But aggressive pecking is normal behaviour, and feather pecking is abnormal behaviour which only occurs in captivity.

Aggressive pecking	Feather pecking
Aimed at the head	Not only to the head but the whole body.
Aimed at a bird that gets under the feet of a higher ranking bird.	Aimed at birds that are quietly eating or taking a dust bath.
Feathers are sometimes pulled out, but are never eaten.	Feathers that have been pulled out are frequently eaten.
It is only a sign of compromised welfare if it happens a great deal.	This behaviour always indicates a problem.

Feather pecking and cannibalism are undesirable behaviour and occur in all husbandry systems.

Difficult to reverse

Feather pecking and cannibalism are signs of reduced bird welfare. Feather pecking leads to higher feed intake and cannibalism leads to losses. Once feather pecking and cannibalism happen in a flock, they are very difficult to eradicate. So prevention is key.

Feather pecking is first visible at the bottom of the back, at the base of the tail. A bare chicken is more susceptible to injury and infections.

Cannibalism is when one bird eats the skin, tissue or organs of other birds, dead or alive. The area around the vent and the abdominal organs are the parts most prone to pecking.

Caged vs floor chickens

Feather pecking occurs in all husbandry systems. It is a bigger problem with floor chickens than with caged birds. Floor kept feather peckers in a large group can create more victims. It is easier to keep chickens in a lower light level in a cage house. Lower light intensity means that the chickens are less active, which also reduces feather pecking.

Bare chickens cost money

Relationship between plumage and feed intake

Feather intake

155

115

0

0   25   50   75   100

% plumage missing

Feather pecking is bad for bird welfare and costs money. A bare chicken eats 20% more food just to keep warm. Another rule of thumb: a chicken needs to eat 4 grams more per day for every 10% of feathers it loses. Bare chickens that move about a lot or go outdoors need even more feed.

Light and feather pecking

Dimming the light reduces feather pecking and cannibalism. Darkening the house makes the birds less active.

Red light helps control cannibalism, although it is not known exactly why. Red light reduces the light intensity and makes the chickens less active. However, red light and steady light intensity can also cause the birds to become more aggressive.

Feather loss

Feather loss can also have other causes.
- Deficiency of amino acids, vitamins, minerals and roughage in the feed.
- Problems with intestinal and general health, so absorption in the intestine changes.
- Mycotoxins in the feed or litter.
- Diseases (e.g. certain skin mites) that affect the feather follicles, making it difficult for new feathers to grow.
- Neck moult: only the neck feathers disappear. In pullets, this is associated with coming into lay too early. In older hens, it is caused by stress from disruption, cold, a change of feed or broodiness. If your pullets are affected, ask your feed supplier whether you can add more protein on a temporary basis.
- Wearing of feathers due to contact with the equipment (cages, drinkers, etc.)
- Wearing because of mounting of roosters (in breeders).

Feather pecking

The first sign of a deficiency in the feed is the disappearance of the feathers that are usually lying around on the ground. Every now and then you will hear a screech of pain when a feather is pulled out. The birds will start to suffer injury and the process will be difficult to stop.

Chickens start feather pecking later if they are fed meat and bone meal in their feed. Tests reveal that feather pecking still happens, but only at an older age than with feed containing vegetable protein.

Causes	Prevention	What to do about it?
● The lack of litter (in the rearing phase) ● Deficiency of fibre, minerals, vitamins or amino acids in the ration ● Chronic gastroenteritis ● Irritation by red mites ● Poor house climate, bright sunlight ● Boredom and stress ● Too high light intensity combined with one of the above factors.	● Ensure a smooth transition from the rearing house to the laying house. You should not suddenly give chickens accustomed to a dark house a lot of light. Keep to the same times for switching the light on and off, feeding routines etc. ● Combat mites. ● Provide distraction in the form of dry and loose litter. Scatter grain or roughage regularly to keep the place attractive. Suspended ropes, aerated concrete blocks, peck blocks, corn cobs, grass etc. Give them something new regularly. ● Feed mash instead of a pellet and enough high-fibre raw materials.	● Check the nutrient levels in the feed. Provide extra vitamins and minerals. ● Dim the light or use a red light. ● If the worst comes to the worst with a flock kept on litter, try using blinders (chicken spectacles). From an animal welfare point of view, this is strongly discouraged.

Signals of feather pecking

Because of the difference in colour between coverts and underfeathers, feather pecking is more noticeable in brown chickens.

Chickens lose feathers every day, which remain on the ground. If the feathers start disappearing from the ground, they are being eaten: a sure sign that there is something amiss in the flock.

For these 11 week old hens, these aerated concreted blocks are a much loved pecking object.

This house has a scattering system suspended above the scratching area. Grain is scattered four times per day, about one-third of the daily ration. The core feed is matched to the scattering grain.

Cannibalism

Causes

- When an egg is laid, part of the vent emerges with it. Birds with too much abdominal fat push their vents out much further.
- Chickens that lay floor eggs literally expose their vents.
- Too much light in the nests. The vent always bulges out a little during laying, forming a target for cannibalism.
- Deficiency in the feed (protein, vitamins or minerals).
- Injured birds are a target.
- No uniformity: chickens that are too light weight are the first victims.

Extra measures (particularly with floor systems) after eliminating the causes

- Every day, remove any weak, frightened, injured and dead chickens from the flock or place them in the sick bay.
- Make sure that the eggs are not too heavy; this causes bloody vents.
- Dim the light or use a red light.
- Provide things to peck at, like aerated concrete blocks and roughage.
- If the cannibalism is feed-related, tell the feed supplier and get them to deliver new feed if necessary.

With roughage the hens are kept busy and it is good for their gut health.

Dead chickens are very interesting to the other birds in the flock. Remove them immediately, as this can encourage cannibalism.

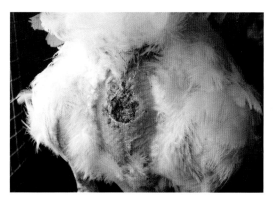

Feather pecking and cannibalism have nothing to do with aggression.

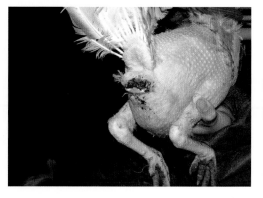

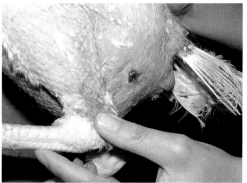

You can recognise cannibalism on dead birds because they will have been scavenged from behind. On live birds with bare patches you will see peck wounds, especially around the vent. This starts off small but can lead to terrible injuries.

CHAPTER 5

Egg signals

A chicken can only produce lots of good quality eggs if she is in top condition. If the eggs are not good this is an indication that there is something wrong with the chicken or her care. Take a good look at the egg size, cracks and breaks, the shell quality and the shell colour. The eggs can show you a lot about whether there is something wrong with the chicken or the system.

Eggs are the reason you keep laying hens. So make sure that the last step in the production process is managed well too.

Look at the eggs whether there are some abnormalities or defects. Be aware that different signals require different actions. Egg signals can be distinguished into the following categories:
- Shell abnormalities resulting from the hen itself
- Shell abnormalities originating before laying
- Shell abnormalities occurring at laying
- Shell defects due to egg transport and egg collection
- Shell abnormalities due to diseases
- Shell defects due to other causes.

Double yolked eggs are mainly laid at the start of the production period. The younger the hens are when they come into lay, the more double yolks are laid.

A fresh egg?

Consumers want a strong, fresh, uniform egg.
There are various ways in which even a consumer can assess the age of a 'fresh' egg.
This can be determines by:

- size of the air cell
- albumen viscosity
- position of the yolk in a boiled egg

The air cell

The air cell is a major factor in a damaged egg, but also in potential contamination of the egg.
It is also a characteristic that tells you about the age of the egg or its storage conditions.
During storage the egg mass gets smaller and the air cell gets bigger. And because the albumen expands when you boil an egg, the air cell in a boiled egg is much smaller than in a raw egg.

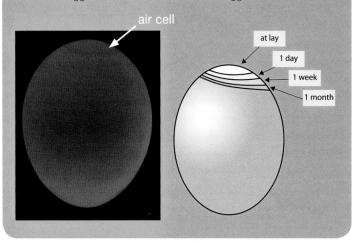

air cell

at lay
1 day
1 week
1 month

boiled

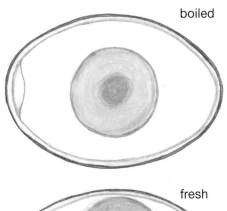

fresh

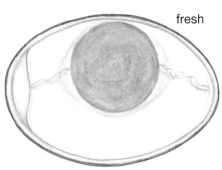

When you open a boiled egg, you'll find the yolk in the middle. But this is not the position in a unboiled egg. Although stabilised by the chalazae, it kind of floats against the shell. As the egg gets older, the chalazae get weaker and albumen gets thinner. So in a old egg that has been boiled, you'll find the yolk against the shell. Also a signal of the age of the egg.

Fresh or old egg?

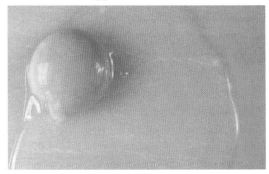

Old egg   *Fresh egg*

Temperature and moisture play a major role during egg storage. Eggs are usually picked up twice a week from the farm. In this case, the best storage temperature is 18°C. If they are stored for more than 10 days, 10-12°C is better. Avoid large temperature fluctuations to prevent condensation.
Condensation promotes the development of microorganisms on the shell (fungal growth) which can penetrate the pores and contaminate the contents. The relative humidity during storage influences moisture loss in the egg (resulting in weight and quality loss). Keep the humidity between 75 and 80%. Poor storage conditions accelerate ageing of the egg.

Internal egg quality

Eggs go straight to consumers from the laying farm without any processing. So the egg quality must be good and the eggs must not contain any impurities. The internal aspects that determine the quality of an egg are flavour, residues, germs, inclusions and albumen viscosity.

Undesirable substances

Eggs must not contain any undesirable substances such as blood or flesh specks, residues of antibiotics, antiparasitics, pesticides, environmental pollutants, or sometimes worms. Destroy any eggs produced during administering medication and the withdrawal period following this. Consumers expect this level of care and honesty, and residues can cause health problems.

Although very rare, a roundworm can sometimes be found in chicken eggs.

Blood and flesh

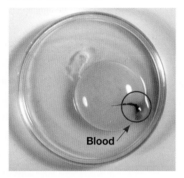

Blood

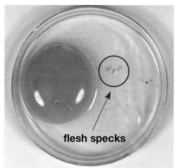

flesh specks

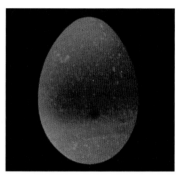

Blood and flesh specks in a whole egg are only visible during candling if they are not too small. Flesh specks are particles released from the oviduct and are not visible when the eggs are candled. Blood specks are the result of the yolk sac being torn at a site with blood vessels or in the oviduct. This can be caused by fright reactions or IB infections. Yolk sacs usually tear at sites with no blood vessels.

LOOK-THINK-ACT

What does a different yolk colour tell you?

The colour of the yolk is mainly influenced by the amount of colouring agent in the feed. If the yolk colour remains too pale compared with the rest of the flock (i.e. uneven colour distribution in this flock), check the following:
- Was these hens' digestion poor due to an infection, for example? This would make the chickens less able to absorb the colouring agent in the feed.
- If you think the chickens were not sick, ask your feed supplier to adjust the feed.

Shell abnormalities caused before laying

External quality criteria of table eggs are weight, colour, shape, and the strength and cleanliness of the shell. You can tell a good deal from the outside of an egg. Cracks and breaks often point to problems with the cage floor or the egg belt. Defects or dirty shells are directly related to the health status of the hens, the composition of the feed and/or soiling of nests or manure on the cage floor.

The shape of an egg can differ considerably as a result of the hereditary tendency of the hen. It has nothing to do with illness or management.

The egg on the left has calcium speckles, which can have different causes.

Ridged top. Can be caused by stress during laying.

Ridged shell; usually caused by infectious bronchitis.

Distress among the hens can also cause 'cracking' of the eggshell while the shell is being formed. These eggs are known as body-checked. The problem occurs when the shell starts to form in the early evening.

Pimples; can have various causes, e.g. infectious bronchitis. It may also be due to the breed of bird.

The shell of this egg is misshapen (slab-sided egg) because there were two eggs in the oviduct at the same time lying together in the shell gland. This is not associated with illness; it is mainly a consequence of the hereditary tendency of this hen.

A membranous egg missing most of its shell. Possible causes: rapid successive ovulations at the start of the laying period if hens come into lay early. The egg is laid before the shell starts to form. The oviduct cannot keep pace with rapid successive yolk production. Membranous eggs and soft-shelled eggs can also be caused by other factors such as too high temperature or illness (e.g. Egg Drop Syndrome).

Pimples; local roughness, usually at the blunt end of the egg. This can be caused by infectious bronchitis. In that case the contents are watery. Please note: it can also depend on the type of bird, but then the shells will be thicker and there will be nothing wrong with the internal quality.

The egg tip is rougher and thinner and shows a clear separation from the healthy part of the shell: glassy tips. Cause: infection by a particular Mycoplasma synoviae strain in the laying organs.

Larger eggs produced at the end of the laying period can have weaker shells. Adjust the calcium content of the feed in good time and provide extra calcium. Make sure the hens feed well before the dark period starts, as shells are mainly deposited at night. There may also be a problem with the hens' feed intake (disease, high temperatures).

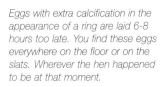

Eggs with extra calcification in the appearance of a ring are laid 6-8 hours too late. You find these eggs everywhere on the floor or on the slats. Wherever the hen happened to be at that moment.

Finding abnormal eggs

It may seem as if you get more abnormal eggs in cage systems. But this is a misconception. In cage systems all the eggs are collected, but in floor housed systems only the eggs that are laid in the nest and in the litter are collected. In floor housed systems, some abnormal and thin-shelled eggs are not laid in the nest, so they are not noticed. So they don't count as abnormal eggs.

In any case membranous eggs (shell-less eggs) are difficult to find. In floor housed systems you will find these eggs in the manure under the slats where the birds perch to sleep. With cage housing, they often don't roll off properly because there are other chickens in the way, so they are often trapped on the floor. So always look under the cages or in the manure pit under the slats.

The shell of this egg is ridged. A possible cause is stress during laying.

Unexpected white eggs

Are you unexpectedly coming across white eggs? This can be caused by a coccidiostat in the feed (nicarbazin). Even traces of this can cause problems. In fertilised eggs it kills the embryos. Another reason is diseases like IB, TRT and ND.

Shell abnormalities caused after laying

Blood on the shell comes from a damaged vent caused by too heavy eggs or vent pecking.

Dust rings are caused by eggs rolling on dirty floors. In cages, dust in the egg drawer can cause dust rings. Also make sure that the shells can dry off before the eggs roll on to the belt. You can do this with an egg saver which ensures that the egg is dry and lands on the belt slowly so dust does not stick to the shell. Of course, the eggs should not remain in the house too long. Clean the egg belts regularly.

The presence of egg yolk on the shell of some eggs is a indication that hens are pecking at eggs and eating them.

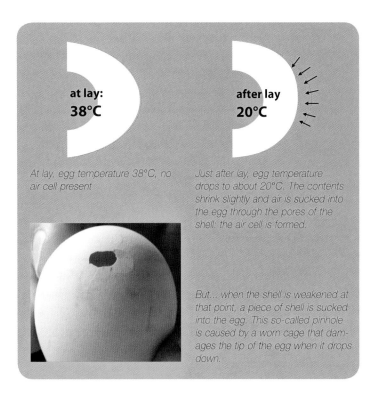

**at lay:
38°C**

**after lay
20°C**

At lay, egg temperature 38°C, no air cell present

Just after lay, egg temperature drops to about 20°C. The contents shrink slightly and air is sucked into the egg through the pores of the shell: the air cell is formed.

But... when the shell is weakened at that point, a piece of shell is sucked into the egg. This so-called pinhole is caused by a worn cage that damages the tip of the egg when it drops down.

Manure on the shell could be the result of intestinal diseases which cause the hens to produce thin manure. Wet droppings can also be caused by incorrect feed composition. Also check the nest expulsion system if you are using roll-away nests. If this is not working properly or it closes too late, eggs can be soiled by dirty nest floors. Even if you are using manually collected nests, make sure the nests are nice and clean.

Laying rhythm

The hen goes to the nest to lay an egg 24-26 hours after the yolk is released from the ovary. If the egg is laid as a membranous egg after as little as 4 hours, the hen is not yet sitting on the nest. And even if the hen has no egg in the oviduct, she will still go and sit on the nest at the usual time. Eggs with extra calcification in the form of a ring are laid as late as 28 hours after the yolk is released (laying is postponed for 4-6 hours). The hen goes and sits on the nest in time, without laying the egg. Later she will simply lay the egg wherever she happens to be at the time, as she did not need to look for a nest at that time. So you will only find these kinds of eggs in cage systems or some of them in the litter.

With brown eggs, these eggs can be recognised by the white ring on the side of the egg. With white eggs you won't notice them because you can't see the white ring on the white shell.

Cracks and breaks

Eggs can get damaged and show breaks, hairline cracks, dents or holes immediately after laying. Look at the site and the nature of the damage. A small hole in the pointed or blunt end indicates that the egg hit the floor too hard when it was laid. This can indicate a worn or too hard wire floor in the cage, or a bare spot in the nest. Cracks and breaks on the side indicate damage while rolling off the cage or nest to the belt or during transportation.

Carefully check the route the egg follows from the hen to the collection table: are the eggs rolling gently enough, are they rolling against something, are the transitions between belts properly aligned? The more eggs there are on the collection belt, the more breakages and cracks you can expect. So make sure you collect the eggs often enough: at least twice per day. Every system has points to watch out for. If 95% of eggs in a floor house arrive at the same place on the egg belt because the hens prefer those particular nests, there will be a greater chance of damage. Allow the egg belt to run a couple of times to spread out the distribution of the eggs.

Check whether there are any specific problem places using an electronic egg (a transparent egg with built-in electronics) or by collecting eggs in certain places and candling them.

Cracks and breaks can also occur if the hens in cage systems are frightened, causing them all to suddenly fly up and jump around. If this happens often, find out what is scaring them, for example a wild bird in the house or an electric current on metal components.

An electronic egg stores information about the movement and shocks an egg is subject to, which you can download later on. This can tell you where eggs can get damaged on a belt and in the packing machine.

If there are too many eggs lying close together, you will see damage on the side of the egg.

Cracks and breaks occur more often at the end of the laying period. The eggs get bigger and the shells weaker towards the end. Or there is a lack of calcium in the feed.

It could also be that the belt is running too fast and keeps getting switched on and off, jolting the eggs against each other. It is better to run it slowly rather than fast with lots of stops.

A wrongly placed egg will lead to broken eggs in the tray. This doesn't only mean loss of the egg, but also contamination of other eggs in the tray and a smell of rotten eggs.

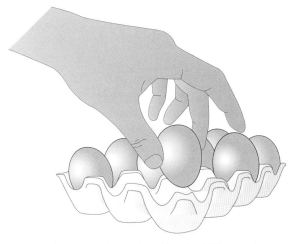

Eggs should be placed point down (air cell up). The air cell spot is most vulnerable and should never carry the weight of an egg during transport. Furthermore, by placing the egg point down the yolk will stay nicely centred inside the egg.

Eggs are often transported on conveyors. These can be rod conveyors or belts with a flat surface. With a rod conveyor debris will fall through, while on flat belts egg handling can be more gentle. Starting and stopping belts creates cracks. Proper adjustment in relation to the automated packer or egg grader is essential.

An egg with a crack or break on the side. The shell was damaged when the egg rolled off. A crack or break at the blunt end indicates that the nest floor is too hard; the egg is damaged by falling on the floor when it is laid.

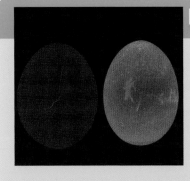

LOOK-THINK-ACT

Recent or old cracks?

Both eggs show cracks, but there is a difference. The cracks in the shell on the right are thicker. These cracks are of a few days old. This is important to know in order to find the cause.

Washing eggs

In some countries eggs are washed before packing them into the final consumer packs. In Europe this is not permitted while in the USA this is obligatory by law. Egg washers spray the eggs with warm water and sometimes chloride. Then eggs are treated with brushes in order to wipe off debris. Finally eggs are rinsed in clean water before they are dried by huge blower systems. As the natural protection layer of the egg (cuticala) will be damaged to some degree, after washing a cooled transport and storage chain is necessary in order to keep the eggs fresh. Unwashed eggs do not need to be cooled during transport and storage; they will remain fresh for many days in ambient temperatures.

If eggs are transported to a central packing station, systems known as 'farm packers' are often used. These systems don't grade eggs into categories but only place them on trays to enable transport on trays with all eggs positioned with the air cell up.

You can achieve a higher hygiene status with plastic trays than by reusing pulp trays. Egg processing can for the most part be automated: pulp trays are less suitable for this. In addition, plastic pallets are becoming more and more widespread. Plastic trays are more hygienic, provided they are properly cleaned.

It is a complete waste of your investment in poultry-farming if the eggs break in the final stage of production. A small investment in pulp or plastic trays is well worthwhile. The advantage of plastic is that it lasts longer, although the investment is higher. Transport in baskets creates unnecessary cracks and breakage (crack percentages up to 20%). On trays these same eggs only show 2% cracking.

Egg packing stations can be organised in two ways: 'inline' and 'offline'. Inline means that an automatic egg grader is directly connected to the conveyors from the chicken houses. Offline means that the egg grader is placed elsewhere and eggs are transported to the packing station on trays. In that case a loader is needed to place the eggs from the trays onto the egg grading machine.

Quality report from the egg wholesaler

As a poultry farmer, you not only have to respond to the signals from the chickens but also to those from your customer: the egg wholesaler. The wholesaler will check your eggs for various criteria. The quality report can tell you where improvements need to be made, such as egg weight, yolk colour and damage. If you keep an ongoing record, you can anticipate developments instead of only taking action when you get an alert from your customer.

Signal	Report parameters	Possible action, now or next flock
Too light, too heavy	Average weight	Amount and type of layer feed
		Choice of breed, hen weight from rearing farm; coming into lay too early or too late
Too light, too dark	Yolk colour	Amount of red and yellow colouring agents in the feed, feed intake, intestinal and general health
Weak shell	Breaking strength	Layer feed, technical guidance, choice of breed or too high a temperature and therefore too low feed intake
Too low	Haugh Units (HU; freshness)	Amount of protein in the feed, protection against IB, general health of the hen
		High storage temperature
Too light/pale	Shell colour (brown eggs only)	Choice of breed in the next cycle
Dirt on eggs (including manure, dust, mites, fly droppings)	Staining	Good, modern laying nest, good layer feed, good bird health
		Speed of the egg belt
Too many cracked eggs	Cracks (star cracks, shell membrane intact)	Egg collection system at the poultry farm (lift or packer), breaking strength too low
Too many hairline cracks	Hairline cracks (invisible to the naked eye, eggs burst open when cooked)	Egg collection system at the poultry farm (lift or packer), breaking strength too low
Too many open breaks	Open break	Egg pecking by hens, rough handling of eggs, breaking strength too low
No stamp, incorrect code, not clearly legible	Egg stamp	Adjust and maintain egg stamping machine

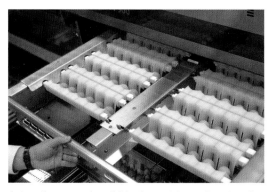

Automatic egg graders divide all eggs into quality categories. The simple systems do this only by weight while the more sophisticated machines also are able to detect cracks, dirty eggs, shell colour and bloodspots inside the eggs. Modern graders can provide statistic and traceability data and also control inkjet and labelling systems. Simple systems allow eggs to accumulate in reservoirs per grade, other systems treat eggs individually in order to prevent collisions and thus breakage of the eggs. It is very important that the machines are well calibrated and that this is checked periodically.

Grading eggs

In egg graders abnormal eggs can be spotted by human operators (candlers) who are able to pick out these 'offgraders' in a darkened candling booth. The most sophisticated machines have automatic detection systems:

Crack detection, by means of acoustic analysis.

Dirt & leaker detection by means of cameras and image analysis.

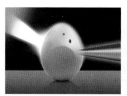

Blood detection by means of spectrum analysis.

Shell colour detection by means of colour sensors.

The output products can be packed in an enormous variety of different consumer packs. Worldwide over 800 different consumer packs are used. An egg grader should be as flexible as possible to enable use of future pack designs.

  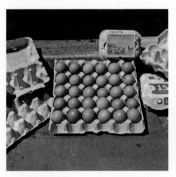 

If you are having production problems, also think about how the birds were reared, as this may be a cause. A non-uniform flock will mean that some birds get off to an uneven production start, resulting in lower peak production.

Always test the blood

Check the initial status of all your flocks at the beginning of the laying period (at 20 to 22 weeks) with a blood test, or freeze blood serum to test for antibodies in the event of later problems.

Egg production problems

Is the number of eggs produced consistently below standard, or is the quality of the eggs poor? In floor housing systems eggs can also be lost because they are not being laid on the nest. It is not only infections that can cause production problems. Production can get under way late because of a poor lighting programme, too little feed or even too much, or a too low body weight, for example. Poor flock uniformity at the end of the rearing period can also result in a low production peak. If production is low, look out for other symptoms as well.

Clinical examination

With many production problems, you will see little if anything wrong with the birds. But it is still important to have a good look at the flock and listen to it. If you can see any other disease symptoms, then it can be much easier to pinpoint a cause. Weigh your birds weekly so that you will notice any changes in their weight in good time.

Blood test

In the event of production problems, take blood samples from 24 birds from all over the house. Do this right at the beginning of the problems (acute stage) and 3 to 4 weeks later. During the three weeks after the infection, the birds will have formed antibodies which will show up in the blood.
The blood samples can be tested for antibodies against IB, EDS, TRT, *Mycoplasma gallisepticum* (Mg), *Mycoplasma synoviae* (Ms) and avian encephalomyelitis (epidemic tremor, AE).

Checkpoints in case of production problems

When investigating production problems at flock level, besides the obvious points, make sure you pay additional attention to the history of the production, the egg quality and the rearing period. In floor housing systems a lot of eggs can also be lost because they are not being laid in the nest. This manifests itself in too low production right from the start. Check for losses due to floor eggs by walking through the whole house a couple of times in the morning when the light is switched on.

History
- Production curve (egg production, production peak, egg mass, persistency of production), mortality percentage, age of birds at 50% production

Egg quality
- Egg weight and uniformity (weight, size)
- Shell quality: shell colour and thickness, cracks and breaks, glassy tips, dirty shells, malformations: rings, asymmetry, pimples
- Internal egg quality: height of thick albumen (Haugh units), albumen and yolk colour
- Distribution of first and second grade eggs, characteristics of second grade eggs

Information on the rearing period
- Body weight and uniformity
- Transition from rearing to production period
- Vaccination schedule
- Autopsy reports (unfortunately, too few autopsies are performed on young hens in the rearing period)
- Lighting programme

The perfect egg is our goal.

Causes of egg production problems

Drop in production (%)	Other signals	Cause	Effect on eggshell	Impact on egg content
0	None	*Mycoplasma synoviae*	Glassy egg tips	-
0-15	Respiratory problems: sometimes slight, sometimes more serious caused by mixed infections	*Mycoplasma gallisepticum*	Possibly determined by additional infection	Possibly determined by additional infection
1-10	3-5% have puffy heads/wet noses	TRT	Pale shells	-
5-15	Reduced feed intake, respiratory complaints, wet manure due to kidney problems	Infectious Bronchitis*	Pale shells, rings, pimples, asymmetric eggs	Watery albumen, loose air cell, broken chalaza
10-90	Egg-binding, kidney problems, respiratory problems during rearing	Infectious Bronchitis during rearing**	Pale, weak shells, shell-less eggs, pimples, rings, asymmetric eggs	Watery albumen, loose air cell, broken chalaza
5-20	Seriously short of breath, blood on nose, death	ILT	-	-
5-20	Poor feed intake	Feed quality	Varying	-
10	Rickets	Vit. D3 deficiency, Ca/P balance	Thin shells, shell-less eggs	-
10-60	No	Avian encephalomyelitis	No	-
30-50	No	Egg Drop Syndrome	Pale, weak shells, shell-less eggs	
10-100	Severely short of breath, diarrhoea, nervous symptoms, high mortality	Avian flu, Newcastle disease	Pale shells, shell-less eggs	Watery albumen, loose air cell

* It is rare for all the signals and symptoms listed here to occur simultaneously in one flock.
** See remarks on IB. The consequences during the laying period are also associated with the age at which the IB infection occurred in the rearing period.

CHAPTER 6

Health

A chicken can only perform optimally if it is healthy. Sick birds usually stop drinking first. So it is very important to keep a close eye on the daily water intake; any reduction is often the first sign that something is wrong.

Sick birds also have less desire to eat even though they need extra energy to boost their natural defences. Producing proteins for growth and egg production therefore takes a back seat as survival is their priority. Consumption of trace elements and vitamins increases in diseased birds.

Don't only call the vet if there is a problem, but also to keep a finger on the pulse and make other arrangements before something goes wrong.

Disease always costs money. Besides the cost of lower production (growth, eggs, development) you also have to pay for the treatments: the direct cost of the drugs used and the indirect cost in case the eggs cannot be sold for consumption due to drug residues.

Unsaleable eggs are one of the costs of disease.

Disease signals

Diseases often manifest themselves in symptoms, and Laying Hens are perfect for picking up on these.

Recognising a disease therefore starts by being able to assess health properly. As soon as you have a good image of a healthy chicken in your mind, it is easier to pick up on subtle changes (see also chapter 1). So start with the disease signal: what does it tell you, and what is causing it? Use your senses: look, listen, smell and feel.

Categorising diseases

As soon as you identify a symptom, categorise it in a main group. This makes it easier to make a diagnosis because you can often rule out a number of diseases.

Signals that can help you identify symptoms are:
- the gastrointestinal system
- the respiratory organs
- the laying organs (egg production)
- the musculoskeletal and nervous system
- the skin and feathering

There are also acute diseases with a high mortality rate.

Although good observation provides you with a lot of information, it is often necessary to run further tests to ensure that the birds don't get the wrong treatment. These include:
- Autopsy by the vet. This can sometimes be done on the farm.
- Also lab tests such as a bacteria culture, virus culture, blood test, tissue test and parasite test.

So it is important to have a good, reliable lab relatively near to the farm.

One size doesn't fit all

The annoying thing is that one symptom doesn't always lead you straight to a particular cause. You need to recognise several signals to be able to identify a disease. Conversely, you may also find that one disease can have various symptoms, so it may not be easy to narrow down. Besides their main symptom, respiratory problems, diseases like avian flu and Newcastle disease can also cause lameness and diarrhoea, so they fit into several different groups.

Symptoms of disease: hunched up, eyes shut, feathers fluffed out.

Disease signals in pictures

This rearing hen is breathing through its open beak because of a respiratory inflammation.

This layer is lying hunched up on the bottom of a cage. The comb is very pale. This hen needs attention.

Sick birds tend to hide.

Biosecurity outside the house

Biosecurity is about keeping poultry healthy. First and foremost, disease must be kept outside the door: external biosecurity. Pay sufficient attention to routes along which infection can take place, and draw up a hygiene plan in which you list all activities and their frequency and order. Aspects that should always be included in the plan are rules for visitors and outside vehicles entering the farm, delivery and removal of animals, removal of sick animals and carcasses, use of boots and clothing, refilling the disinfection tank, cleaning, disinfection and pest control.

An effective layout

A well-designed layout with a properly separated clean and dirty area is the basis for good external biosecurity. Only allow clean vehicles to enter the premises via the clean route. Enable unloading to take place via the dirty route, for example for feed trucks, so that they do not have to drive up to the house. Germs are very easily transmitted on shoes, clothes and hands and are then transferred to the hair and the airways. People who have to enter the house must always do so via the compulsory decontamination area.

Prevent young hens from bringing red mites into the laying farm by switching on the main lighting half an hour before capturing them. The lice will then leave the chicken. Just before capturing the chickens, switch the main lighting off and the blue replacement lighting on.

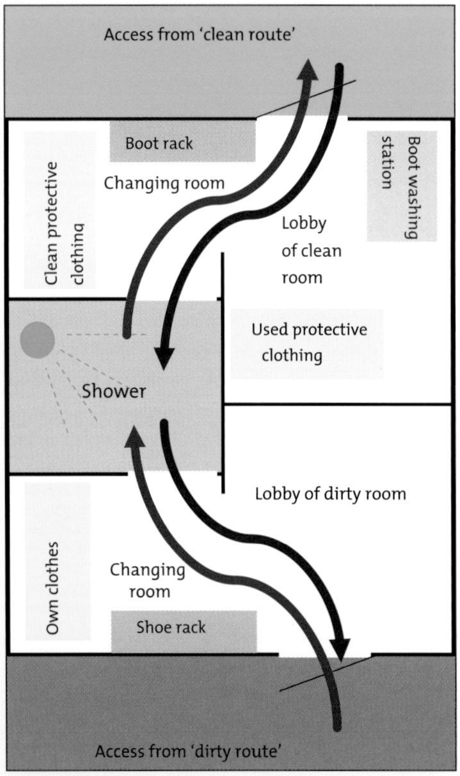

Decontamination area

The only access to the house must be through a decontamination area. In the clean area, visitors wash their hands or shower, if facilities exist. They then put on clean protective clothing and boots and can then proceed into the house. When the visitors leave the house, the boots are removed and cleaned. The visitors remove the protective clothing, change back into their own clothes and leave the decontamination area on the 'dirty side'.
Visitors must not be permitted to take personal items into the house unless absolutely necessary. And a mobile phone is definitely not necessary.

Diagram labels:
- Access from 'clean route'
- Boot rack
- Changing room
- Clean protective clothing
- Boot washing station
- Lobby of clean room
- Used protective clothing
- Shower
- Lobby of dirty room
- Own clothes
- Changing room
- Shoe rack
- Access from 'dirty route'

Ideal situation: *room with two doors, separate place for dirty and clean clothing and boots, possibly showers*

Bad situation: *dirty and clean people crossing over the same bench, and dirty and clean clothes hanging on the same hooks.*

Biosecurity in the house

Also avoid spreading diseases in the house itself: internal biosecurity. Take a critical look at your own working methods. How clean are your overalls and boots? There are potential pathogens in the manure and dust you carry around with you. What route do you take when you walk round the house? Always walk from the area with the youngest birds to the area with the oldest ones. Do you always make sure you go from clean to dirty, and do you wash your hands and change your clothing or boots when you go to a different section or house? Don't allow affected or sick birds to run around: always remove them. And never put any birds you remove in a sick bay in the house. They must be killed and disposed of carefully or sent for autopsy.

What is your vermin situation? Vermin are a major transmitter of diseases. So it is essential to protect against and control vermin. The level of biosecurity you should aim for partly depends on the husbandry system you use. Very strict measures are pretty pointless in free range systems, for example. It is vital not to just follow the rules every now and again but to do so consistently ALL THE TIME.

Don't forget to check your fan outlets. Make sure they don't blow straight into another house.

Remove dead birds

Remove dead birds daily and do not allow them to pile up next to the house door. Place them in a refrigerated carcass storage facility or remove them immediately. In floor housing systems check the laying nests, slats and litter every day. Make sure everything is easy to see and easily accessible.

Dead animals are a breeding ground for bacteria. Hens peck at dead birds, spreading bacteria fast among the flock.

Controlling mice and rats

Mice and rats are wary transmitters of disease, including salmonella and pasteurella:

- Seal holes in floors and walls and cracks or chinks in buildings, windows and doors.
- Remove feed, manure and egg residues wherever possible and tidy up rubbish promptly.
- Don't use anterooms or lofts as storage spaces.
- Store things in closed rooms.

It is impossible to keep a farm entirely free of rats and mice. So remain alert and check the house regularly for traces of rats and mice. Set up mousetraps and bait boxes with poison. Ideally you should bring in a specialist company to perform regular and effective vermin control.

Make sure that there are no places where mice can hide, for example by keeping the first few metres around the house free from vegetation. Grass is fine, but keep it short.

Controlling flies

Make sure that the manure remains sufficiently dry (> 45% dry matter) as fly larvae thrive in wet manure. In systems with a manure belt, run the manure belt regularly to remove the manure from the poultry house, particularly in the hot season. Always install UV fly killers in the house and count the adult flies that land on it at set times. If the number increases noticeably, there is a breeding ground with larvae and maggots somewhere.
Find the source and treat it immediately with larvae killer. Besides chemical agents, these days you can also use a biological method with predator flies.

Flies spread germs when they are eaten by the chickens, via hair and in faeces.

What seems to be the trouble?

When assessing a flock, there are three key questions:

1. What am I observing? Observing is not just looking, but also listening, smelling and feeling. The symptoms, a disease profile.
2. What has caused it? The diagnosis.
3. What can I do about it? What action do I need to take? Treatment and prevention.

Also ask yourself whether the problem is in the whole flock or just in an individual bird.

How to perform a clinical examination

Step 1. Medical history
- What are the symptoms?
- When did the problems start?
- Do you have the results of any previous examinations?
- Have these problems been going on for some time on the farm, or have they occurred before?
- What are the technical characteristics?

Step 2. Looking at the birds
- When assessing several houses or flocks, always look at the healthy ones first, followed by the sick ones, and then look from young to old.
- Assess the flock as a whole: are there noticeable symptoms? If so, how many animals show symptoms and how serious are these? How is the uniformity, the distribution in the poultry house and behaviour? What can you tell, for example, by the plumage, feathers in the barn and digestion of the manure?
- Perform a physical (clinical) examination of any abnormal or sick birds.
- Assess the external and internal egg quality.

Step 3. Further examination
What additional examinations are needed to establish the cause? Take the time to select the right birds for this examination and send in enough birds and/or samples. Possibilities are:
- Autopsy of sick birds and follow-up examination of organ samples (bacteriological, virological, parasite). Depending on the problem, your vet may tell you to send in sick birds, and in some cases healthy ones too, for autopsy, or he/she may perform an autopsy on the farm.
- Examination of paired blood samples (samples taken on appearance of symptoms and three weeks later)
- Examination of other sample material such as windpipe swabs and manure
- Examination of the house climate.

Step 4. Additional laboratory tests.
These may sometimes be necessary in order to obtain a definitive diagnosis.

Step 5. Recommendations for other and subsequent flocks.

Gastrointestinal problems

In birds, the vent, or cloaca (Latin for 'sewer'), is not only the place where solid waste is excreted, but also waste products from the kidneys. Hens and roosters also excrete something else here: the egg and sperm respectively.

Three different types of manure/waste products can be distinguished:

1. Normal intestinal droppings, the voluminous ones which are often comma-shaped. When normal, the surface is covered in very small cracks and the droppings remain dry when squeezed.

2. Caecal droppings. In the mornings chickens deposit a sticky, damp, shiny pile which ranges from caramel to chocolate brown in colour.

3. Urates from the kidneys. Birds do not urinate like mammals (they have no bladder) but convert their urine into uric acid crystals which are deposited in a white layer on the droppings.

Besides abnormal manure, there are other general indications of gastrointestinal problems: huddling together, fluffed up feathers, lethargy and death. Birds with digestive problems have too little energy and therefore have a greater need for warmth. Increase the house temperature for a while. Chronic digestive problems can lead to deficiencies of proteins, vitamins, minerals and trace elements.

Left: intestinal droppings with caecal droppings on top, right: caecal droppings

Abnormal droppings and possible causes

Signal	Possible cause
Homogenously thin	Intestinal problem
Pool of water with strings of urates and clumps of droppings	Some virus infections (such as Gumboro and renal IB)
Feed components visible	Poor digestion
Orange-red, sticky strings	Too long without food, or intestine affected by e.g. coccidiosis
Fresh blood in stools	Possibilities include coccidiosis
Dark green droppings	Loss of appetite or severe acute diarrhoea with undigested bile salts
Thin yellow caecal droppings with gas formation	Intestinal dysfunction or incorrect feeding
Watery, white droppings	Kidney problem or inadequate feeding resulting from infection

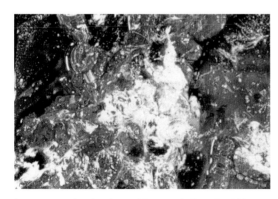

Large amounts of watery, white excreta from the kidneys: renal IB or Gumboro

Fresh blood in the droppings comes from the intestinal tract and indicates acute caecal coccidiosis.

Respiratory diseases

Chickens suffering from respiratory diseases will be short of breath and often breathe through their open beaks. But this can also be an indication of problems with the house climate, fever, pain and anaemia.

Signals of respiratory problems

Signals that are specific to respiratory problems:

- Unusual breathing noise: sniffing, snuffling and snorting, rattling or hawking, crying, yawning and screeching. The best time to observe this is when the chickens are at rest (e.g. in the evening when it is dark);

Many respiratory problems start with a slight inflammation of the eye membrane, which can be recognised by slight foaming in the corner of the eye.

Short of breath, but no unusual noises. These chickens have a fungal infection of the lungs.

Action needed: this hen has a serious eye membrane inflammation, and its sinuses below the eyes are swollen.

- Shortness of breath: animals breathe through the open beak and make pumping movements with the abdominal muscles;
- Inflammation of the eye membrane (wet or thick eyes), nasal cavity and pharynx;
- Enlarged head due to swelling of the sinuses.

There are also some less specific signals: sitting huddled together, fluffed up feathers, lethargy and death.

Signals of climate problems

If the house is too hot, in floor housing systems birds will look for cooler spots; they will sit huddled together along the walls, for example. They will often sit with beaks open and necks stretched. Their wings will hang loose against the body and their tail will bob up and down. But you won't hear any noise. The comb and wattles are dark red. Birds at risk of suffocating lie on the ground with feet pointing backwards and necks stretched. Chickens that are cold will huddle together in groups, often with feathers fluffed up and heads withdrawn. They will often look sick.

Signals of fever

The normal body temperature of a fully grown chicken varies between 40.6 and 41.7°C. Chickens can develop a fever from bacterial and viral infections. This is most noticeable with Gumboro and IB. Because the birds are sick, they will huddle together so they can't bring down their temperature. They will then die from overheating and their feet will lie stiff against their bodies. If you open one up just after it has died, steam will escape: their body temperature may have increased to as much as 45°C!

Pain signals

Sitting with the beak open can also be a sign of severe pain in a bird. Adult hens express pain much less clearly.

Anaemia signals

Chickens sometimes seem to have difficulty breathing because of a respiratory problem, but on closer examination they may be found to be suffering from severe anaemia caused by a serious lice infection, for example.

Breathing problems

In the early stages of respiratory problems, the signals are often the same and there is nothing to tell you whether the problem is mild or a serious, possibly notifiable disease. You can tell a lot from the other signals associated with respiratory problems. A poultry farmer should therefore always talk to his vet straight away if mortality increases or if production, feeding or drinking dwindles. Further laboratory tests will be needed to confirm the illness.

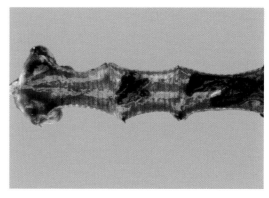

Red windpipe caused by infectious laryngotracheitis (ILT)

Sounds as a signal

Type of sound	Cause	Possible cause
No sound, beak open	No excessive mucus or inflammatory fluid in the airways	Fever, high house temperature, fungal infection on lungs or pain
Sniffing	Slight mucous membrane irritation with small amount of inflammatory fluid, moist eyes	Poor house climate: high ammonia, low relative humidity. Vaccination reaction or start of viral infection.
Snuffling and snorting	Mucous membrane irritation in upper airways, sometimes coupled with eye membrane inflammation	Viral or bacterial infection, vaccination reaction
Rattling or hawking	Mucous membrane irritation in nasal cavity and upper windpipe with excessive mucus formation	Poor house climate with *E.coli*. If symptoms appear suddenly: IB or NCD
Squealing, yawning and screeching	Inflammation of the airways with stiff mucus, often sudden death from suffocating	AI, NCD, IB or ILT with *E. coli*

Clap your hands once or whistle loudly when you enter the house. The chickens will stop in their tracks and you will be able to hear soft rattling and coughing (flock level).

If you suspect an airway infection, hold up a chicken with the breast to your ear and listen and feel for any rattling or other unusual breathing (chicken level).

Disorders in the locomotive organs

The causes of locomotive problems can be traced back to the nervous system (brain and nerves) or the musculoskeletal system (muscles, bones, joints). The former leads to lameness, wryneck and compulsive movements, such as with avian encephalomyelitis, vitamin E deficiency, Marek, AI and NCD or bacterial meningitis (often a hatchery infection). The neurological form of Marek's disease mainly affects white hens and usually starts at around 6 weeks.

Limping in one or both legs?

In limping birds, watch for symmetrical or asymmetrical limping. Asymmetrical limping can be caused by foot damage, a joint inflammation or Marek's disease. Symmetrical limping often indicates a more central cause such as Reovirus-induced tenosynovitis or bone pain.

Compared with the meat sector, foot problems such as Reovirus-induced tenosynovitis, perosis and hock tendon problems occur relatively infrequently in the laying sector.

Bone pain from rapid osteoporosis can occur at any age. Specialist knowledge is needed to be able to interpret the degree of deterioration of the bone.

If all birds in a group of young chicks are lame in the same foot and they have been vaccinated there against Marek's disease, this may indicate a badly performed vaccination. In brown hens arthritis is sometimes found with orange discolouration: amyloidosis. This is often associated with hatchery infections during Marek vaccination.

Footpad ulcers

Footpad ulcers mainly occur with broilers, but in floor housing systems they can also affect layers if the litter is very wet. In the first 14 days the skin on the feet is still thin; it only becomes callous later on. Uric acid and ammonia in a wet poultry house affects the skin, causing cracking and inflammation. If the litter is wet when the bird is older, the footpads can become severely inflamed. If the footpads remain dry and clean for the first few weeks, the birds will be much less prone to footpad inflammation when they are older, even if the litter is in poor condition. In cage housing footpad ulcers can be caused by sharp wire; a particular risk in new cages.

The chicken on the left is lame in one foot: asymmetrical.

Footpad ulcer on a laying hen.

Locomotion disorder caused by a Reovirus-induced tenosynovitis infection.

Sudden increase in mortality

An increase in bird mortality, particularly if it happens suddenly, is always a major cause for alarm. Both the welfare of the flock and its production and therefore your financial result are seriously affected.

Be aware of when the alarm bells should start ringing. More than 0.1% mortality per day can be described as a significant increase. If the mortality rate is more than 0.5% per day, you are talking about a dramatic increase.

Removing and processing dead birds

Seriously sick and dead birds must be removed from the house as quickly as possible to minimise the risk to the others.

Dead birds should ideally stored at a temperature of max. 7°C outside the house and must be covered up to keep flies away until they can be removed from the farm. This helps minimise germ growth and unpleasant smells.

When losses occur, always check the surviving birds.

A bird incinerator is a hygienic way of processing dead birds.

Naturally, you must NEVER allow dogs, cats, rodents and insects anywhere near dead animals. Dead animals can be removed in three ways:

1. A special cadaver truck picks the birds up and takes them to a specialist destruction company. These cadaver trucks should be parked as far away from the poultry houses as possible.
2. The dead birds are incinerated in an incinerator on the farm at least once per day.
3. Composting is a good method if it is managed and performed properly. This method is mainly used for birds kept on litter, but there is also some experience with birds in cages.

Which method you choose depends on the type of farm you have and your local laws and regulations.

Possible causes

Infectious causes (germs)	Non-infectious causes
● Botulism	● power failure (ventilation failure)
● acute coccidiosis	
● E. coli infection	● house climate related (heat stress; carbon monoxide poisoning; ventilation failure)
● Erysipelothrix rhusiopathiae	
● Avian Influenza (AI)	
● Newcastle disease (ND)	
● Gumboro disease (during rearing)	● stampeding/fright reaction
	● water and feed disruptions
● necrotic enteritis (Clostridium perfringens)	● toxicity (salt)
● Salmonella gallinarum	
● Salmonella enteritidis (during rearing)	
● fowl cholera (Pasteurella multocida)	

Always hand over carcasses on the public road. The cadaver truck must not enter your 'clean' route. Carcasses must be chilled down to approximately 7°C in carcass bins to avoid unpleasant odours and the spread of germs.

Summary of the main diseases

A disease is relevant to commercial poultry if it occurs frequently and if an outbreak results in economic loss. Zoonoses - infections that can be transmitted from animal to man, such as Salmonella, avian influenza (fowl pest) and Erysipelas - are extremely important in terms of public health.

Viruses

Infectious bronchitis (IB)

The IB virus is very common. The first problems arise as early as 2 to 3 days after the virus has hit a flock, and it takes no time at all for the entire flock to be infected. Infected birds spread the IB virus for many weeks after recovery, and the virus can survive for months in the intestines. As soon as a symptom-free carrier is stressed, for example after transportation or other infections, the IB virus can re-emerge.
In young birds the IB virus results in respiratory problems, often followed by an *E. coli* infection. In layers you may often only notice production problems. In some cases the IB virus even causes severe kidney complaints with a high mortality rate and also digestive problems.

What symptoms occur depends on the type of IB virus and the climate inside the house.
The IB virus changes continuously, producing new variants all the time. An effective IB vaccination programme should therefore be based on thorough lab research. Unfortunately, this is often 'forgotten.'

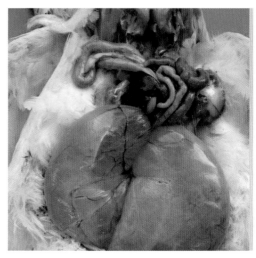

A false layer looks normal on the outside but can't lay an egg because her oviduct is deformed. The bird adopts a penguin stance. This abnormality, which we see in birds in production, is caused by an IB infection early in the rearing period.

Infectious laryngotracheitis (ILT)

ILT is a viral disease which causes severe breathing problems, loss of production and death. Brown layers in particular display severe symptoms: serious breathing difficulties with a nasal discharge, sometimes containing blood.
In the event of an ILT outbreak you can give an emergency vaccine, which can limit the damage considerably.
ILT symptoms can be caused by a vaccine virus spreading from an older flock.

Breathing difficulties

Egg Drop Syndrome (EDS)

EDS occurs occasionally, particularly in the first few months of production and with heavier breeds (brown layers). Vaccination keeps the number of outbreaks limited. There is no treatment for EDS because it is caused by a virus. The disease spreads slowly: more quickly in floor systems than in caged systems. Typical of the infection are the large number of weak shells and membranous eggs.

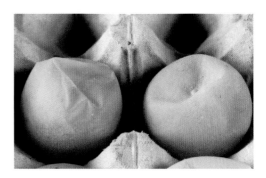

Membranous egg caused by Egg Drop Syndrome

Viruses

Marek's disease

Marek's disease is a Herpes virus infection that causes tumours. It resides in the feather follicle cells and is therefore also found in dust particles from skin and feathers. Feather dust - and therefore often Marek's disease as well - can be found everywhere where chickens are or have been. It is therefore important to vaccinate chicks immediately after hatching and house them in a clean environment. You should also keep different aged chicks separate. If inadequately protected chicks are infected in the first six to eight weeks of life, disease symptoms will emerge from about 15 weeks.

The disease has three forms:

Neurological: often asymmetrical lameness in the feet, for example. Marek nervous form can be found from 6 weeks onwards.

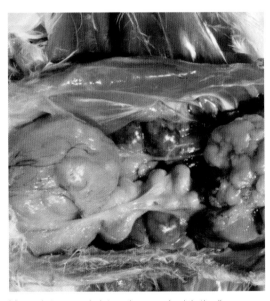

Visceral: tumours in internal organs (mainly the liver, spleen, ovaries and sometimes other organs). This form is the most common and may result in extremely high mortality.

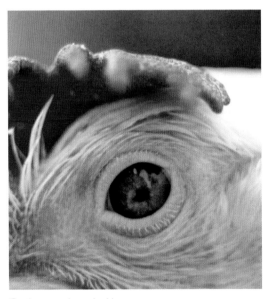

Ocular: grey, irregular iris

Viruses

Newcastle disease, ND (pseudo-fowl pest)
ND is not called pseudo-fowl pest for nothing. The symptoms of ND and AI are alike as two peas in a pod. ND is also notifiable and requires the same measures to be taken as for an AI outbreak. The ND virus is a paramyxovirus, while the AI virus is an orthomyxovirus. In various countries there is a legal requirement for commercial poultry to be vaccinated against this.
In the event of a ND outbreak, a rapid ring vaccination in the area around the contaminated farm will limit the risk of it spreading further.

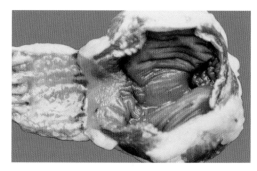

Proventricular bleeding

Avian influenza (AI, bird flu or fowl pest)
Symptoms after infection with a high-pathogenic form of the AI virus are a sudden increase in mortality that suddenly rises again within a few days; swelling of the head, purple discolouration of the head, comb and wattles; subcutaneous haemorrhages; shortness of breath; diarrhoea; lameness; huddling and ruffled feathers. The AI virus can therefore cause all kinds of symptoms, making it difficult to identify immediately. AI, NCD, ILT, TRT and sometimes IB can all produce similar symptoms. AI is notifiable.
In every outbreak of disease with sudden extremely high mortality, you should suspect AI.
But there are also some AI strains that cause only very mild symptoms in poultry, or none at all.

AI, bird flu

Gumboro disease
Gumboro is a serious viral disease that affects young chickens and can be accompanied by sudden death, typical watery, yellowy-white urate diarrhoea and reduced immunity. Rearing pullets are much more susceptible than broiler chicks, with mortality rates sometimes exceeding 50%. Autopsy reveals typical abnormalities: the bursa is swollen and surrounded by a glassy skin (oedema), often with haemorrhaging. Muscle haemorrhaging and swollen kidneys complete the picture.
There is no treatment. Vaccinating young chicks can effectively prevent the disease, so it is very important to determine the right time to vaccinate. To do this, a good, fast lab test of the chicks' existing immunity levels is needed. The importance of this is often underestimated. Some Gumboro vaccines can be given in the hatchery.

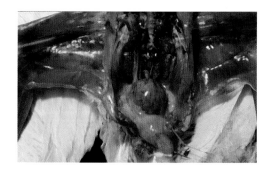

Inflamed bursa

Avian encephalomyelitis
Avian encephalomyelitis is a viral disease that causes lameness in young chicks, a significant drop in egg production of between 10 and 15% in layers and an approximately 5% drop in breeding results in parent birds. The virus is transmitted to progeny through the egg. In pullets, around 15% (or anything up to 60%) display lameness (squatting, lying on the side) and tremors. The mortality rate among affected birds is 50%. The tremors are usually only seen or felt in a small number of birds by picking them up.
Vaccinating parent flocks gives almost 100% protection to their progeny.

Typical lateral position caused by avian encephalomyelitis

Bacteria

E. coli or *peritonitis*

This is very common in adult hens: moderately to severely increased losses of birds in top condition are often the only symptom. Production is usually more or less unaffected. Sick birds are rare, and production is usually maintained. In a flock of young chicks with colibacillosis, the birds will huddle together with raised feathers. Their breathing is laboured, they snort and cough and often produce thin droppings. There may be lame birds, and some birds may stop producing. Losses are between 0.2 and 1% per day. Autopsy reveals inflamed air sacs, liver capsule and heart sac. Risk factors for colibacillosis are viral infections of the airways, poor house climate and inadequate hygiene.

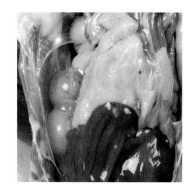 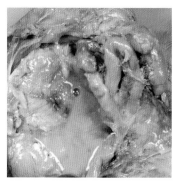

Normal abdominal cavity   *Peritonitis in a layer*

Salmonellosis

There are a great number of Salmonella serotypes, some of which can cause illness in poultry. Typical symptoms are diarrhoea and greatly increased mortality, particularly in the first and second week of life. Losses from peritoneal inflammation caused by *S. enteritidis* are sometimes observed in adult layers. *S. gallinarum* results in widespread mortality, including in older birds. Autopsy reveals inflammation of internal organs including the ovary and peritoneum.

Antibacterial treatments and vaccinations are ultimately ineffective.

There is a simple way of eradicating Salmonella, so it is necessary to develop a specific farm-based approach for each case. In humans, some serotypes that come from poultry, including *S. enteritidis* and *S. typhimurium*, cause food poisoning with severe diarrhoea.

Abnormal ovary with follicle stalk formation, commonly seen with an S. gallinarum infection   *Dark, inflamed pericardium and swollen liver caused by S. enteritidis infection*

Brachyspira infection

Brachyspira bacteria cause avian intestinal spirochaetosis (AIS), a chronic intestinal inflammation resulting in reduced nutrient intake. This causes deficiency and lower resistance. The symptoms include frequent drops in production, diarrhoea, weight loss and higher bird losses.

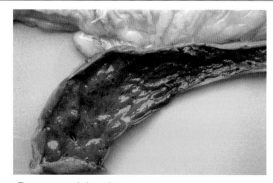

Foamy caecal droppings

Erysipelas

Erysipelas is caused by the bacterium *Erysipelothrix rhusiopathiae* and is particularly relevant to turkey and free range chicken farms. The bacteria can survive for many years in the ground, but also in carcasses.

Besides slightly lethargic, weak chickens, thin manure, higher mortality and a significant drop in production, there are very few typical symptoms.

The next flock can be vaccinated as a preventive measure. Warning: Erysipelas is a zoonosis! Humans can be infected through skin wounds, producing local inflammation with a red, painful swelling.

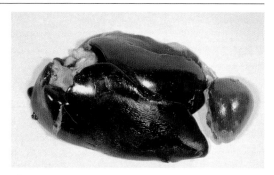

Swollen liver and spleen caused by Erysipelas

Bacteria

Mycoplasmosis

Mycoplasma gallisepticum (Mg) infection mainly causes inflammation of the breathing organs in chicks. Older birds also suffer from production problems. The infection is a chronic lifelong condition.

Swollen head caused by mycoplasmosis

Fowl cholera (*Pasteurella multocida*)

Fowl cholera has an acute and a chronic form and is caused by the *Pasteurella multocida* bacterium. The acute form is associated mainly with higher mortality and diarrhoea. The chronic form can result in inflammation of the comb and wattles. Fowl cholera can be a serious problem in layers, turkeys and ducks, and the bacteria are also found in other wild birds and even in rats, mice and pigs. Options are preventative vaccination and antibiotic treatment.

Swollen comb and wattles caused by Pasteurella

Coryza (*Avibacterium paragallinarum*)

Coryza is caused by the bacterium *Haemophilus paragallinarum*, and mainly occurs in hot areas. It mainly affects older birds. It is especially seen in multi-age farms that are never depopulated. Morbidity is high but mortality is low if uncomplicated, although it may be up to 20%. The bacteria survive 2-3 days outside the bird but are easily killed by heat, drying and disinfectants.

This usually acute, sometimes chronic, highly infectious disease is characterised by catarrhal inflammation of the upper respiratory tract, especially nasal and sinus mucosae.

1-3 days after the first contact there is a rapid onset of disease over a 2-3 day period with the whole flock affected within 10 days, resulting in increased culling. Carriers are important with transmission via exudates and by direct contact. It is not egg transmitted.

Symptoms
- Facial swelling
- Purulent ocular and nasal discharge
- Swollen wattles
- Sneezing
- Dyspnoea
- Loss in condition
- Drop in egg production of 10-40%
- Loss of appetite

Prevention is obtained by stocking coryza-free birds on an all-in/all-out production policy.
Vaccines (so-called bacterins) can be used; at least two doses are required. Commercial bacterins may not fully protect against all field strains but reduce the severity of reactions. Live attenuated strains have been used but are more risky. Controlled exposure has also been practised.
Vaccines are used in areas of high incidence. Birds recovered from the challenge of one serotype are resistant to others, while bacterins only protect against homologous strains.

Cause unknown

Chronic enteritis

Chronic enteritis often occurs at around 25 weeks when production is increasing rapidly. The first symptoms are distress, thin manure, reduced feed intake, poor production increase, messy plumage and loss of feathers which are subsequently eaten. Autopsy reveals a clear intestinal disorder in the first 20-30 cm of the intestine (the duodenum), i.e. too much and too thin content with an abnormal colour. There may also often be small necrotic lesions caused by Clostridium spp. These lesions are dark-grey in colour, sometimes with obvious bleeding. The irritated intestinal mucous membrane tries to repair itself by renewing more rapidly but is then unable to mature properly.

This causes digestive problems and changes the composition of the gut contents. The changed gut contents form a breeding ground for bacteria that do not belong in the first part of the gut and lead to further irritation, making the enteritis chronic. Chronic enteritis is best treated by improving the gut function and suppressing the irritating gut bacteria. This can partly be controlled via the feed and with copper sulphate.

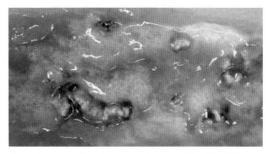

Open intestine showing necrosis caused by Clostridium.

Untidy plumage: signal of chronic enteritis

Internal parasites

Histomoniasis

Like *Eimeria, Histomonas meleagridis* is a single-celled parasite that can cause severe disease symptoms and death in chickens and turkeys. Autopsy reveals typical inflammations of the liver and caecum. In most countries there is no registered treatment, but chemotherapeutic products are used in some countries. However, you can tackle the intermediate host (the caecal worm), which reduces the pressure of infection.

Histomonas causes typical lesions in the liver.

There are five species of *Eimeria* with relevance to chickens. Each species has its own favourite spot in the gut

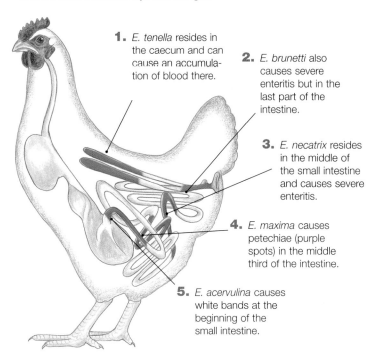

1. *E. tenella* resides in the caecum and can cause an accumulation of blood there.

2. *E. brunetti* also causes severe enteritis but in the last part of the intestine.

3. *E. necatrix* resides in the middle of the small intestine and causes severe enteritis.

4. *E. maxima* causes petechiae (purple spots) in the middle third of the intestine.

5. *E. acervulina* causes white bands at the beginning of the small intestine.

Coccidiosis

Coccidiosis is caused by several types of *Eimeria*. This single-cell gut parasite is common in poultry and causes very minor to severe harm to the gut. Moderate infections can result in subclinical disease, while mass infection can lead to clinical symptoms such as diarrhoea and sometimes death. Coccidia spread as oocysts which are difficult to destroy with disinfectants. That is why almost every farm in the world is infected with coccidiosis. And this is one of the most harmful diseases in poultry farming.

The severity of the symptoms depends on the species of *Eimeria*: from slightly retarded growth with *E. acervulina* infection to sudden death with *E. necatrix* or *E. brunetti* infection. The manure is often abnormal. With caecal coccidiosis (*E. tenella*) there may be fresh blood in the droppings. Coccidiosis can also lead to bacterial enteritis (Clostridium).

Prevention

Good hygiene and good litter quality are essential for limiting harm from coccidiosis. Vaccination is possible with live vaccines: attenuated (low-virulence) or non-attenuated (wild type) strains. Such a vaccination is common with birds intended for reproduction, and for layers in alternative systems. Most vaccines contain live coccidiostat-susceptible strains.

Pale bird suffering from caecal coccidiosis

The economic consequences are often underestimated: only the tip of the iceberg (clinical coccidiosis) is visible, but below that there lurks a much greater risk (subclinical coccidiosis).

Worms

To get a good impression of the worm infection in a flock, it is recommended to have a worm egg count performed every six weeks. To do this, take a mixed sample of 20 piles of intestinal droppings and 20 piles of caecal droppings.

Caecal droppings are sometimes mixed with the intestinal droppings, but if it is important to differentiate between the parasitic roundworm *Ascaridia galli* and nematode parasite *Heterakis gallinarum*, collect the two kinds separately. The caecal worm resides in the caecum and the large roundworm in the small intestine. The droppings need to be as fresh as possible. The samples should be kept chilled and examined within a week.

Symptoms

- Usually a slow process, i.e. a chronic condition
- Slight diarrhoea, weight loss or retarded growth
- Hens 'dry out': comb shrinks, production ceases
- With persistent, severe infection: pale comb and wattles, exhaustion
- The disease generally becomes more severe in young birds than in older ones.

Post-mortem

If the worm egg count cannot distinguish between the types of worm, post-mortem examination of some birds may be required:

- to establish whether the worm is an *Ascaridia galli* or a *Heterakis gallinarum*
- to rule out other infections with similar symptoms
- to determine the severity of the infection and the damage to the intestine.

Treatment

There are various ways of tackling worm problems:

- deworming every six weeks to prevent a serious threat of infection. Every three weeks for caecal worm/tapeworm.
- dropping analysis every six weeks, post-mortem examination in case of doubt; treatment on that basis.
- only deworm if infections are found by chance.

Most common worms

Large roundworm (*Ascaridia galli*)
Large roundworm infections are often symptom-free. Death only occurs with severe infection, usually as a result of intestinal damage by larvae or blockage of the intestine. The severe symptoms are mainly seen around three weeks after infection.

Caecal worm (*Heterakis gallinarum*)
The caecal worm is not actually, or only very slightly, pathogenic but it can transmit the serious disease blackhead (histomonas) via the eggs. For treatment advice for worms, you will need to know whether they are large roundworms or caecal worms. The difference can only be ascertained by autopsy or by collecting intestinal and caecal droppings separately. You should do this if there is histomonas in the vicinity, for example.

Tapeworm (*Raillietina*)
The tapeworm is easy to recognise by its jointed structure. This worm damages the intestines. When a worm segment containing eggs is excreted via the manure, the eggs are eaten by beetles (including litter beetles) and ants. Chickens reinfect themselves by picking up these intermediate hosts. After about two weeks, more segments containing eggs are excreted and the cycle starts again.

Worm eggs in the droppings. What now?

Depending on the type of worm and the number of faecal eggs per gram (EPG), it may be necessary to de-worm the chickens:

- EPG >1000 of the large roundworm
- EPG > 10 of the hairworm
- in case of caecal worm infection, de-worming is not immediately necessary.

Also look at the production figures and the condition and health of the birds.

Severe large roundworm infection

Red mites

Red mites can carry harmful bacteria or viruses. The bloodsucking parasite transfers them as it passes from one chicken to the next. With a severe red mite infection, chickens lose a lot of blood which can lead to anaemia, and in turn death. An average red mite infection reduces the chicken's resistance, but there are few external symptoms other than the fact that their plumage is rougher because they pick at the itchy skin. Chickens control red mites naturally by taking regular dust baths.

Accumulation of live mites on the egg belt: an ideal means of transport for eggs and mites.

Fighting the mite in an empty house

It is best to control red mites when the poultry house is empty. Clean it well to remove the hiding places such as under piles of manure. If you use pesticides, make sure you follow the instructions properly. A too low temperature can dramatically reduce the effectiveness of certain products. Heat treating the house can also greatly reduce an infection. Birds' nests on the outside of the house are a potential source of infection: remove them.

Pest control during an inspection

During an inspection, you need to act fast at places where the population manifests itself first. You can treat locally straight away without having to treat the entire house, for example with silica powder or biodiesel. Another possibility is physiological control. This is a method in which the birds' blood is made unattractive to the red mite by feeding the chickens vitamin B2 or garlic. However, this is less effective with a severe infection. You need to start early.

Only mite-free hens

Make sure you obtain mite-free pullets if you have the choice.

LOOK-THINK-ACT

What does this behaviour tell you?

These hens are dust bathing. This can be seen by the stretched-out leg of the hen in the middle. Dust baths help to fight red mites. However, if you find the hens resting in the litter, it could indicate that they are looking for other resting and sleeping places than perches. Red mites like to hide under perches. As the risk of a red mite infection increases, the chickens start to stay away from the perches. In that case the number of floor eggs will probably also be increased (mites in nests).

Host

Every parasite needs a host to survive. The red mite prefers various species of wild birds and the chicken. You will find it regularly in birds' nests. The red mite can also occasionally attack mammals (such as rodents, dogs and cats) and even humans. In optimum conditions an egg will develop into a fully grown red mite in seven days. The red mite sucks blood even in the nymph stage. During daytime red mites hide in small holes everywhere in the house equipment. They only move onto the hens in the dark. This means that if you check your animals during daytime, you will not find mites, while at the same time you might have a big problem.

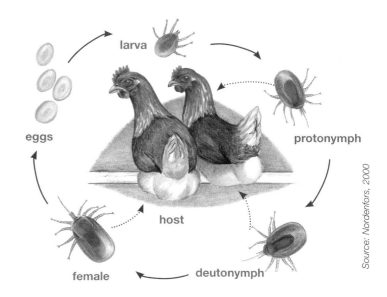

Source: Nordenfors, 2000

Do the red mite test yourself

Red mites can be tracked down by knocking on the system or by scraping in cracks with a knife while holding a white sheet of paper underneath. You can also hang up mite traps in which the mites will hide. Inspect places where red mites are likely to hide daily. In cages these are mainly under the egg protection panel, fastening clips and gutter. In floor housing systems you will find them under perches, under slats, in dry droppings and in laying nests.

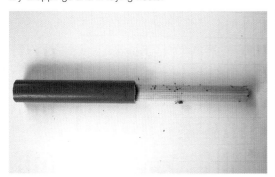

Mite trap. Create a hiding place for red mites. Insert a rod in a dark tube and hang it underneath the perch. After 24 hours, remove the rod and check for red mites.

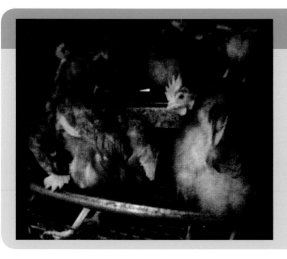

LOOK-THINK-ACT

Also carry out some covert surveillance during the dark period.

This photo was taken at night with an infrared camera. The chickens should be sleeping peacefully. But as you can see, the hen on the left is picking at itself to stop the itching. This is disturbing other chickens as well.

Index